THE MONSTER IN THE MACHINE

THE MONSTER IN THE MA-CHINE

Magic, Medicine, and the Marvelous in the Time

of the Scientific Revolution

ZAKIYA HANAFI

DUKE UNIVERSITY PRESS   *Durham and London 2000*

© 2000 Duke University Press
All rights reserved
Printed in the United States of America on acid-free paper ∞
Typeset in Aldus by Keystone Typesetting, Inc.
Library of Congress Cataloging-in-Publication Data
appear on the last printed page of this book.

To my mother

WINIFRED ELIZABETH HARRIS

and to my grandmother

DOROTHY DIXON WALL

(1900–1989)

great lovers & stewards of books

CONTENTS

PREFACE

A monster is whatever we are *not,* so as monsters change form so do we, by implication. The human and the monster vie for space between two thresholds of transformation: the upper limits are godhood, the lower limits are bestiality. We stake out the boundaries of our humanity by delineating the boundaries of the monstrous, whether by defining the criminal, the insane, or the merely inhumane. This book attempts to track the ways human beings were defined in contrast to supernatural and demonic creatures during the period of the Scientific Revolution. The advent of nonmagical or secular ways of looking at the world and its inhabitants turned the human being into a sort of creature quite distinct from animals and gods. Since the monster partakes of both animality and divinity, I have used monstrosity as a contrasting parameter in order to note the shifting place of humans in a newly secular world.

Given the transformation that took place in seventeenth-century Europe—call it "decontextualization of the world" if you are a historian of magic, or posit an enigmatic transition that cannot be explained, regardless of its etiology, given the accelerated process of desacralization—what happened to the category of the monstrous when its association with the sacred became no longer credible? One answer to this question is that the sacred monster disappeared from the formaldehyde jars of the museum shelves, where its scientific counterpart became safely contained, in order to take up residence in the hidden recesses of the human heart. It was in the mirror, placed in museum collections along with scientific instruments and moving machines, that the scientist caught a reflection of the portentous monster once associated with terror and pollution, with presage and chaos. It is important to note, however, that the monster within became visible only by means of man's own creations: by means of his own machines and his own new sciences. During the period 1550 to 1750 in Italy the sacred characteristics attributed to monsters shifted to products of our own technologies. The monster became a machine. Chapters 2 and 3 establish this hypothesis.

As we read some of the authors of the seventeenth century, we shall see that this shift was not uniform. It was not shared by everybody and was not spread out evenly over regions or periods. Not everybody believed the same things then, any more than they do now. Tides of thought presenting opposing beliefs persisted and mingled with many archaic and seemingly anachronistic beliefs framed in an animistic or even pantheistic cosmogony. Tommaso Campanella was not the only philosopher to subscribe to a conception of a fully sensate universe, peopled by angels and demons who frequently communicated with humans in all sorts of contexts and circumstances.

Chapter 4 traces the effects of new mechanical sciences on medical models of the human body. If *machinelike* and *monstrous* are synonymous in this period, then what was the effect of mechanization on the way we understood our bodies and our diseases, both mental and physical? The ancient precepts of physiognomy not only indicated how a man's character could be read from his features but also taught that inside us lurk a range of bestial attributes. But we are the masters of our humanity: for every defect of beauty or character, the ancient medicine offered a cure based on sympathies and antipathies that linked all things by their corresponding qualities. The new medicine, in contrast, had no cure for the monster within because it denied the capacity to affect one's defective body by curing one's sinful soul. Just as the nonhuman, or the monstrous, became associated with autonomously powered machines, so the human body, likened to a machine, became monstrous in its own eyes.

I argue that Giovanni Borelli's application of physics to medicine was the first true mechanization of the human body. By a paradoxical twist of history, the Italian "macchina del corpo" avoided seeing itself as monstrous because it never admitted a purely materialist explanation for the movements of the passions. Body and soul were materially joined through the functioning of animal spirits, yet their source of animation as they coursed through the blood could never be a purely mechanical one. This was in direct response to René Descartes's physiology that figured the human body as a wholly mechanical automaton.

The advent of the mechanized body indicated a fundamental shift in Western thought that occurred in Naples with Borelli's work on iatrophysics. Another Neapolitan, Giambattista Vico, was the only major thinker in Italy who clearly understood the negative moral implications of Cartesian automatism. He felt it was vitally important to defend the existence of the soul as an autonomous entity since the human spirit can only assert itself through developing a sense of free will. Vico believed that when we view ourselves as machines mysteriously endowed with consciousness we transmogrify ourselves. In chapter 5 I adopt a somewhat unorthodox methodological approach to Vico's corpus: I examine his own health and constitution in order to piece together his medical culture. Through a reading of his autobiography and occasional pieces I fill in a picture of the physiology of the *New Science.*

Chapter 6, which concludes the book, focuses on the category of the monstrous in debates on the proper use of figures in literary language. In a society in which witticisms had the power to paralyze the body and the judgment, it was important to codify their usage. Here the monster represents a call to the antisocial instinct. Metaphors are monstrous because they impair reasoning and lure us into a pleasurable state of indolence.

This is not a definitive or authoritative history of monstrosity in early modern Italy.[1] It is a guided tour designed to stimulate and entertain readers who are interested in a variety of topics and issues: historians of early modern science, of monstrosity, of Baroque aesthetics, of art; literary critics of the *seicento;* scholars of Vico; philosophers of technology, and so on. It is a point of conjuncture for all readers who would meditate, like me, on the advent of the postmodern condition, and who enjoy doing so by looking to the pre-modern and early modern periods for the challenging metaphors that those cultures provide.

The methodological principle that I adopt is that of "seeing the connections" (to use Ludwig Wittgenstein's term) rather than "making explanations."[2] Explanations justify what *is* on the basis of what *was;* connections create new possibilities of inquiry on the basis of what *may* be. Nothing could be more obvious to a Baroque theoretician of wit than that it is our metaphoric capacity, our ability to freely associate, that hones our intellects and makes us into truly creative beings. It can be liberating to allow one's method to define one's object of inquiry, instead of the other way around. Each of these chapters, then, is meant to round up a number of connections and texts in order to illuminate some aspect of monstrosity that is relevant, both for historical and philosophical reasons, to our present time.

My selection of sources might seem eclectic to the specialist reader. I chose authors and texts for the aptness of their language or the relevance of their thoughts to my stated themes. Since many of the texts are untranslated and unavailable in modern editions, it was necessary to cite at length some authors. The most important goal for me was to allow the early modern voices to be heard, and in fact one of the "service-oriented" goals of this book is to make these texts available in modern English. All translations into English are my own unless otherwise noted.[3]

It has been my desire to give full voice to the people of the period I was studying. The ultimate effect should be that of a mosaic of texts and ideas that offers a colorful, suggestive picture of monstrosity and humanity during this crucial period of Italian (and Western) history. By splicing together this patchwork of voices, by allowing them to speak while providing a richly resonating cultural background, I hope to create the impression of a guided tour through a Baroque museum of ideas.

A similar caveat must be issued regarding my use of secondary, contemporary sources. Authors who are well represented in the bibliography and whose work was seminal to my understanding of the topic (such as Paula Findlen, for example) may not be taken up directly in the text. This should not be interpreted as in any way related to their value. Due to length considerations, I had to make choices as to how I would fill my limited space, and I believed it was more important to make largely undiscussed primary texts available to the general reader than to engage in a specialist's debate.

Something also needs to be said here regarding the geographical and chronological parameters of this work. My narrative begins with ancient Greek and Roman texts, passes to Renaissance Florence, includes important French and English authors such as Descartes and Hobbes, remains mostly on the Italian peninsula in the seventeenth century and then focuses on Naples and on Vico, who died in 1744. The last chapter returns to the seventeenth century and moves out of the history of science altogether into the history of rhetoric. This nomadic eye might seem undisciplined to some readers and exciting to others. My choice of geographical and chronological focus was dictated by the ideas that I attempt to explore in each chapter.

The last issue I would like to mention is the relation between structure and history, the opposition between system versus event, synchronic constellations versus diachronic dispersals, stability versus change, or however one chooses to characterize these opposing terms. By characterizing the "monster" as a universal, cross-cultural entity that undergoes transformations in time yet remains recognizable through changes, to what extent can I argue that monstrosity must be historicized from the outset? How does my own historicization of this term interact with the historicizations that authors of teratology perform in their turn? In other words, if monstrosity is a nonhistorical issue, a philosophical preoccupation that can be found in all societies in any period, what constitutes its specificity such that it can be usefully embedded in some historical process? And is this process that it illuminates significant, meaningful, or wholly a product of my own interpretative act?

I see both the historian and the anthropologist as interpreters moving in time, interacting with their objects of investigation and being changed in the process. This makes our historicization part of our history; our history becomes the same thing as the historicization we make of it. The particular historical event that we designate as such immediately becomes a cultural fact in a cultural context. There is really no mediation possible between historical events, fixed values, and our interpretations of both. The best one can do, perhaps, is to be aware of one's shifting place in the history that one is in the process of creating.

The way I have defined a monster, as a set of characteristics that is constituted according to the cultural order of a particular time, goes some way in responding to this challenge. Perhaps the most succinct way of expressing this interaction between histories, structures, and interpretations is also the most striking and insightful. It comes from the pen of Marshall Sahlins, a historical anthropologist. Remarks Sahlins: "Different cultures, different historicities."[4]

ACKNOWLEDGMENTS

The making of this book has spanned a decade of my life. Many people have helped me over the years and encouraged me to see this project through. Without them, there is no doubt that I would have given up long ago. I would like to express my gratitude to my friends and family, colleagues and students, critics and readers, both known and unknown to me, for having believed in me and "my monster book." Thank you.

This research idea first took form at Stanford University, where Sepp Gumbrecht acted as early midwife, along with Jeffrey Schnapp, René Girard, Carolyn Springer, and Robert Harrison. I also had the great fortune to receive close readings and detailed responses on parts or wholes of earlier versions from Hayden White, Manuela Bertone, and especially John Freccero and Paula Findlen. Naturally, any remaining defects are entirely my own responsibility.

I am indebted to librarian-scholars Mary-Jane Parrine and Sandra Kroupa for sharing their research expertise with me, and to the special collections staffs at Berkeley, Stanford, University of Washington, Cornell, and the Marciana in Venice, for making long hours of transcribing and ferreting a pleasant and rewarding experience.

My graduate and undergraduate students at the University of Washington in Seattle allowed me to expound at length on some untried ideas; their enthusiasm and creativity were a constant source of joy to me. I'd especially like to mention Christine Ristaino and Maneesha Patel, for loving the Italian Baroque as much as I do.

Maria Rosa Menocal, Kika Bomer, Staci Simpson, Fatima Tavares, Laura Cronin, and Shereen and Ramez Hanafi have all provided inspiration and much needed hand-holding when times got tough.

And I am grateful to have found an extraordinary editor in J. Reynolds Smith, whose tenacity and intelligence have made it possible for this work to appear in print.

ONE

The Origins of Monsters

Paris, 1826: an astonished audience at the Academy of Sciences listens to the curious deductions of renowned anatomist Étienne Geoffroy Saint-Hilaire, called in to solve the identity of a mysterious corpse found in an Egyptian sarcophagus. "An accomplished archeologist from Trieste," relates Dr. Ernest Martin some fifty years later, "had arrived in Paris with a rich collection of antiquities taken from digs that he himself had directed in the Egyptian ruins and tombs. Among the objects in this collection . . . there was a mummy, and near it, a terra-cotta amulet, a crude but faithful representation of a monkey, whose huddled posture was the same as the mummy's—one that was customarily given to animals of this species—and from which it was concluded that the bandages concealed the body of a monkey." The fact that the corpse had been found in the necropolis of Hermopolis, a burial place reserved for sacred animals, confirmed this hypothesis.

But when Saint-Hilaire removed its wrappings, he found neither an animal nor a human, but rather a monster; that is, a human baby displaying simianlike deformations. "As soon as the bandages were removed," continues Martin, "he recognized that what he had in front of him was a human being and a monstrous one at that. . . . Thus, a creature born of a woman had been embalmed and then buried, but was viewed as having animal origins." The nineteenth-century scientists concluded that "[the mummy] had been honored as a sacred animal; although excluded from human tombs, it was nevertheless welcomed into the necropolis of Hermopolis."[1]

Honored as a sacred animal, yet excluded from human tombs: we should not be surprised that a little monstrous mummy could provoke such confusion. If the ancient Egyptians, like the Indians and the Persians, felt awe and respect when faced with such an anomalous creature as an anencephalus (born without a brain), many other cultures, primarily in the West, considered such beings abominations. The Law of the Twelve Tables, for instance, was explicit: "It shall be permitted for monstrous infants to be killed immediately!"[2]

In Sparta, Athens, and Rome, deformed births were regarded as sinister presages: the infants were thrown into the Tiber or hurtled over a cliff; the Etruscans threw them in the ocean, or incinerated them; the Longobards drowned or suffocated them; the ancient Germans slew them with a sword or burned them. The people of Rome could even go so far as to stone the mother. Her monstrous parturition was a sign of the heavens assigning her the responsibility of some wrongdoing.[3] If the birth happened to coincide with an adverse event, they demanded her death, but "this anger did not last long, and once it had been appeased, the unfortunate woman rejoined her domestic foyer and anxiously set herself to ablutions to purify herself of the defilement to which the monstrous birth was a fateful testimony."[4]

Evidence for the reaction of ancient and traditional cultures to monstrous births is vast.[5] Whether viewed as an evil omen, a mark of defilement for some grave transgression, or, as the Chaldeans and Etruscans viewed them, as presages to be decoded regarding some future good or bad event, monsters have always been considered highly charged with meaning. And that meaning is a sign of the being's extra-ordinary status.

At the lowest threshold, what is monstrous is simply radically other, nothing more than "nonhuman," a convenient binary opposition, in the same way barbarian, derived from the Greek barbaros, meant nothing more than "not one of us," "foreign," mimicking the senselessness of foreign speech, bar-bar-bar-bar. . . .

A monster is "not human," then, and explicitly signals its foreign status with its body: too many limbs, or not enough, or not in the right place. Monsters are ugly because they are de-formed, literally "out of shape," deviating from the beauty of standardized corporeal order. I know I am human because I am not that. The monster serves to erect the limits of the human at both its "lower" and "upper" thresholds: half-animal or half-god, what is other is monstrous.

Another fundamental meaning of the monster—perhaps the most important aspect for an anthropological understanding of its mythological and social significance—is its hybrid character. Monsters create confusion and horror because they appear to combine animal elements with human ones; they posit a possibility of animal origins, of bestiality, of some kind of "promiscuous" coupling, as Giambattista Vico would describe it. In the jumbled limbs and motley order of its body, the monster threatens to destabilize all order, to break down all hierarchies. Monsters stink of the feral and the forest, of that space outside the law.

Is that why the monster became associated with the voice of God, as an indication of divine will? Ancient mentality reasoned perhaps that, as events happening outside the ordinary course of nature, monsters must come from another world, a different world from the everyday reality, that other world beyond the purely physical, where the gods dwell. *Monstrum* and *teratos*, the Latin and Greek roots of *monster* did not signify a deformed being, but a *sign* in the same category as *portentum, prodigium,* and *ostentum,* terms belonging to the divinatory sciences, only migrating later through association to the natural sciences.[6]

A *monstrum* (from *monere*, to warn or threaten)[7] was by definition a terrible prodigy, not for what it was in actuality—a piteously deformed infant destined to die quickly either by natural causes or by ritual sacrifice—but for what it foretold. A sign of coming calamity, the monster first and primarily was a messenger from the other world. So if the barbarian was distinguished by making no sense, or nonsense, the monster, on the contrary, was distinguished by making several senses: by providing an oppositional corporeal limit to human definition; by eroding the strong conceptual differentiation between man and beast, man and demon, or man and god, pointing to pollution, transgression, a breakdown in social order; and by bearing a sign of warning from the forces of the sacred.

How that sign was interpreted is quite irrelevant, or rather, the interpretation is purely a matter of historical context. The ancient Chaldeans assigned one-to-one correspondences between limbs and exact events, either auspicious or ominous portents: an extra finger meant abundant crops and so forth.[8] More recent "readings" of monstrous bodies, in addition to predicting political changes or calamitous wars, had precise propagandistic purposes. Interpretation of monsters in the Renaissance was nothing less than an alternative political science, more popular and contemporary in nature than the erudite fare of princely counselors like Machiavelli.[9] An appearance of a monster was thus tied from the beginning to an interpretive community, to a social order to which it was addressed and to a priestly cast which was needed in order to decipher its precise significance.

The immolation of those marked by difference was not a practice restricted to ancient or "primitive" cultures, however. In 1543 in Avignon, King Francis ordered a woman to be burned along with her dog because she had given birth to an infant with canine features.[10] As late as 1825 in Sicily an anencephalic baby girl provoked such terror that those attending the birth threw her down a deep, dry well. She was saved only by order of the mayor.[11]

Other examples of the horror provoked by monsters abound, but the reaction of nineteenth-century Western scientists is no less interesting. In their view, the organic reality of a deformed infant gave rise not to terror but to passionate scientific interest, focused especially on the tax-onomical scaffolding required to construct a strong lexical foundation for a systematic categorization of birth defects. Science was called on to re-store order to and extricate the parts from this inextricable chaos of teratological hybrids that combine fantasy with reality and body parts with inappropriate forms. It accomplished this by coming up with precise names for each deformity, by setting up an immense lexical apparatus, an iron-clad taxonomy based on rules and laws of corporeal organization.

Symmelus with tripodia, dicephalus dipus dibrachius, spondylodymus, heterotypic heteradelphia, these are some of the diagnostical labels we use today. But what "dicephalus dipus dibrachius" *means*, for example, is simply "two heads, two feet, two arms." In this distancing through lan-guage, especially by using the scholarly, authoritative language of Latin, we can see how transparent modern medicine's efforts have been to grant normalcy to something that obviously resists classification. Whether real, represented, or used as figures of speech, monsters function as represen-tations of the other face of humanity, some bestial or demonic alter ego that must be repudiated and effaced in order for the authentically human being to assert its civilized selfhood.

The Disappearance of the Sacred Monster

A decentralization of human consciousness took place in Europe in the seventeenth century. Human beings went from a privileged position on the ladder of beings to a uniquely lonely position in a more mechanistic, materialist cosmos during this period. Human priority was displaced not only by the gradual switch from a geocentric to a heliocentric universe. Giordano Bruno went so far as to cultivate the awareness that there is no center at all: he imagined an open universe, populated by infinite worlds.[12]

Mechanistic paradigms, God the Master Watchmaker, Galileo Galilei's mathematization of natural laws, the Cartesian projection of a grid onto space, a space made up exclusively of matter and extension: these factors must have had a disturbing, unsettling effect on a creature who had been accustomed to regarding himself as essentially, ontologically different from the rest of the created world. A comforting solution could be had if everything could be characterized as Art, including human inventions, natural phenomena, and even human beings themselves. At least a kind of ontological continuum could be preserved.

Following this logic, it made sense that, as Thomas Browne expressed it, "there are no grotesques in Nature."[13] Since Nature was God's artifact, He would hardly commit the error of including deformity in it. In fact, following Pliny's lead, Nature's diversity and occasional deviation from its rules of production provided simply more cause for wonder and appreciation. Monsters were seen to bespeak the endless fecundity and creativity of God's handiwork. Their ancient associations with divination, transgression, pollution, breakdown in social hierarchy, terror, sacrifice, in short, their sacred links, seemed to be completely severed during this crucial period.

The analogical network of resemblances between the human microcosm and the worldly macrocosm also ceased to resonate. When the great "Stair or manifest Scale of creatures, rising not disorderly, or in confusion, but with a comely method and proportion"[14] came tumbling down, Man the Western Male Philosopher-Scientist burrowed deeper into his *studiolo* and transformed his world into a theater to be observed. At the center of his museum, in a newly ordered taxonomic space, the seventeenth-century collector could become a removed spectator of his visual encyclopedia. As the creator of a *theatrum naturae,* or theater of nature, the scientist could provide a unifying and exhaustive description of the place of all things in relation to each other and to the whole, with the exception of one creature: himself, standing at the center. "The World was made to be inhabited by Beasts, but studied and contemplated by Man."[15]

Monsters serving as amusing decorations and grotesques became garden ornaments in the late-Renaissance version of the modern amusement park. The welcoming words inscribed on the forehead of the marine monster in Bomarzo invited those who "are wandering the world desirous to see great and stupendous wonders, come here, where there are the horrendous faces of elephants, ogres and dragons."[16] Terror and wonder were the watchwords of a ludic greed for distraction that spilled over into scientific collecting, turning didactic collections into cabinets of curiosities. It also invaded theories of rhetoric, turning the decorative into the main principle of elaboration. It worked its way into social philosophy, turning other cultures into pretexts for exotic operas. Art and nature, *naturalia et mirabilia,* began to merge with alarming insouciance: "In brief," explained Browne, "all things are artificial; for Nature is the Art of God."[17]

This play with the boundaries of the natural and the artificial is one of the defining characteristics of the Baroque period in art history. The Baroque is famous for its exuberance, its opening, spiraling movements, its love of trickery and deception, its extravagantly frivolous evasion of the here-and-now. In our modern studies of Baroque aesthetics, though, we tend to forget sometimes that this immodest hunger for deceptive appearances and transportation into another, super-natural realm was fueled by uncertainty regarding this world order. Specifically, by an uncertainty regarding the limits of the natural itself; and even more specifically, regarding the limits of the "merely human." The celebrated Baroque sensibility for movement, for expansion out of confined spaces, for capturing the instant of metamorphosis, along with renewed interest in Greek mythology and Ovidian stories—this sensibility, we might say, pointed to the dwindling possibilities for transcendence portended by nascent scientificist thought.

The Founding Texts

Monsters do exist whenever people mention them or describe them, even if they may not exist in the real world. Sometimes it is difficult to tell a "real" monster from an imagined one. For this reason it is no more relevant to ask of Aristotle, Pliny, or Augustine whether their monsters were facts or fictions than it is today to ask a producer of a documentary on the Loch Ness Monster. The existence of the film itself attests to the Loch Ness Monster's reality as a creature of our cultural imagination. Monsters exist in all states of mind, from the deeply religious to the caustically secular.

There are monsters, there are accounts of monsters, and then there are compilations and histories of accounts of monsters. There are theories on the generation of monsters, and a science of monstrosity—or teratology—and there are also histories of teratology. Most monsters exist by dint of being repeatedly described in words rather than by being sighted in the flesh.

The literature proliferates according to several principles: the compiler, looking back to preceding times and authors, and around to other languages and countries, brings together secondhand reports, leaves out accounts that strike him as improbable or uninteresting, and adds personal observations and contemporary anecdotes. The speculative literature invariably gives a short history of previous thinkers' theories before launching on a new (or not so new) version of the causes of the generation of monsters or their symbolic meaning. More recent writings on monstrosity adopt a thematic organizing principle, such as "the monster in Western art."

The historians of histories of monsters generally peruse a fixed set of texts, giving an impression of a stable narrative with well-defined genres, and establishing a teratological canon. Elements remain constant, the variations limited, both in what a monster is (definitions inherent to and stabilized by etymological considerations), how it comes to be, what its significance is or is not, and what the genesis and culture of monstrosity have been in Western literatures. In other words, the category of the monster has been defined by an ancient and continuous literature that perpetuates itself in a limited number of discursive contexts which will be described below.

The way I am characterizing this body of knowledge and how it has been constituted is not surprising. History—most naively described as the accounts people give of their past—being what it is, we would expect a certain degree of conformity. There is an element of the real, after all, that inspires writings about monsters: deformed fetuses, anomalous births such as twins, animals with humanoid parts, humans with animal-like features, peoples that have different colored skin or unusual customs, all these provide the stuff of monsters. And that reality, transposed into written accounts, takes on another sort of ontological stability. That is to say, over time and in a culture that reveres book authority, the writings about these "real" phenomena acquire an equal reality.

There is a steady persistence in the themes discussed throughout pagan and Christian writers: Are monsters human beings? Do they have a legal right to participate in the polis? Do they have a soul, and if so how many? Do they foretell a good or bad event? Are they a punishment from the divine forces, or a presage of punishment? Are they an index of the creative powers of Nature, of God? Or are they evidence of Nature's impotence, sin, or error? How are they generated? How should they be classified? There is also a remarkable uniformity of responses to these themes, expressed in the form of a limited number of genres that cross over and borrow from each other. Writings about writings, then, furnish much of the monster literature.

The "teratological traditions" are said by historians to be three in number, based on four prototypical ancient texts: Aristotle's *On the Generation of Animals* founds the "scientific" tradition; Cicero's *De divinatione* founds the "prodigy" tradition; and Pliny's *Naturalis Historia* and Augustine's *City of God* together found the "wonders of nature (or God)" tradition.

NATURAL PHILOSOPHY: ARISTOTLE

The Aristotelian tradition takes the scientific attitude toward monsters as facts of nature, evacuated of any portentous signification, neither shocking nor terrifying to behold. Strictly speaking, the monster represents nothing, shows nothing but the fact that nature can be frustrated in its end. Like a grammarian who occasionally makes a mistake in writing, or a doctor who sometimes prescribes the wrong dosage, nature, in having to work through the brute necessity of matter which resists it, sometimes fails in its purpose. Monstrosities are thus most succinctly "failures in the purposive effort."[18] Aristotle views monstrosity as a graduated scale of imperfection falling away from the realization of the intended perfect form: that of man. Since the particular end of nature in animals is that *like produce like,* "even he who does not resemble his parents is already in a certain sense a monstrosity." The birth of a female, following this logic, is "a first departure" from the intended perfection, but a natural necessity for the continuation of the species.

As for the incidental causes of monsters, the etiology is to be found in the struggle between the formal agent—male seed—to dominate the female matter. If the movement of the seed prevails over the generative secretion of the female (the menses) "the movement imparted by the male will make the form of the embryo in the likeness of itself." Depending on the seeds' strength, heat, abundance, or deficiency the embryo will resemble the father or mother, or pass in resemblance to a remoter ancestor in the paternal or maternal line. Cases of disproportionate bodies, such as satyrism, "in which the face appears like that of a satyr," are due to the fact that "that which is acted on escapes and is not mastered by the semen, either through deficiency of power in the concocting and moving agent [the semen] or because what should be concocted and formed into distinct parts [the material] is too cold and in too great quantity."

When resemblance becomes attenuated even further, all "that remains is that common to the race, i.e. it is a human being." In this gradation of deviation from the perfect male, monstrosity is simply the final term: "If the movements imparted by the semen are resolved and the material contributed by the mother is not controlled by them, at last there remains the most general substratum, that is to say the animal."

Aristotle dismisses as metaphoric usage the popular misconception that monstrous infants are part beast. In any case, different periods of gestation make couplings between humans and sheep, dogs, and oxen absolutely impossible. The same struggle for domination between male form and female matter serves to explain monsters with multiplied parts "so that they are born with many feet or many heads" as well as those who are "born defective in any part, for monstrosity is also a kind of deficiency."[19]

What stands out from Aristotle's writings about monstrosity is the apparent rigorously materialist cast of his argument along with his solid refusal to assign either sacred or bestial origins to monstrous births.

DIVINATION: CICERO

The second tradition stems from a more ancient meaning of *terata* and *monstrum:* terms in the divinatory sciences that mean "sign." By an interesting irony, the text (*De Divinatione*) that represents the portentous aspect of monsters is actually a dialogue written to attack the Stoic doctrine of divination as it was presented by Posidonius (c. 135–50 B.C.). The debate takes place between Quintus, Cicero's brother, who supports Stoic beliefs in the efficacy and need for divination (which includes natural arts, such as dream interpretation and prophecies; and artificial arts involving statistical correlations such as reading of entrails, lightning, and interpretation of events outside the normal course of nature), and Marcus, thought to represent Cicero's views. That is, he is skeptical yet remains within the Stoic paradigms of cosmic sympathy and the organic unity of the universe.

The arguments in favor of divination, though, while based on the assumption that "the whole universe is one,"[20] have little of the supernatural about them. Quintus argues for "a relatively easy science of technical predictions."[21] Since "from the beginning the world was designed in such a way that certain events would follow certain signs, signs in internal organs, in birds, in lightning, in portents, in stars, in dream visions, in the utterances of madmen,"[22] knowledge of the future can be, and should be, obtained through careful observation and record keeping. This activity is commendable "even without any influence, any impulse from the gods, because the frequency of records indicates what happens as a result of something else and what it means."[23]

We cannot help noticing the rigorously semiotic and scientific character of classical divinatory arts: careful observation, record keeping, causal logic based on statistically probable correlations. These texts will be misread and turned into their opposites first by Renaissance Neo-Platonists, who often associated divination with magical arts, and later by Enlightenment thinkers who viewed divination as a superstitious response to the vagaries of fortune, a way of controlling anxiety about the future.

The history of this prodigy tradition is traced out in Livy's and Tacitus's histories, in which the constant correlation between the appearance of portents and public calamity is recorded assiduously. As Cesare Taruffi notes, "this example of recording extraordinary physical phenomena was later followed by other pagan and Christian historians,"[24] until in the fourth century A.D. Julius Obsequens collected together in chronological order all the portents reported by Livy (including those from now lost books of his history) in his *Prodigiorum Liber*. Obsequens's collection of prodigies was then revised and reprinted by Conradus Lycosthenes as *De Prodigiis* in 1552, exemplary of a highly successful and popular genre that the eminent French historian of monstrosity Jean Céard refers to as constituting "L'Age d'or des prodiges" (mid-sixteenth century).

In the modern age, portent divination, including the reading of monstrous bodies for information about the future, reached its apogee of popularity during Reformation times and was used extensively by both Luther and his opponents.[25]

WONDERS OF NATURE: PLINY AND AUGUSTINE

The third teratological tradition is represented by Pliny the Elder, "who would have it that monsters and prodigies are wonders, miracles, that attest to . . . the force of nature and the careful attention that she pays to the destiny of men, or, as Saint Augustine puts it, the power of an artist God who is concerned with perpetually awakening in us a sense of wonder."[26] The pertinent part of Pliny's monumental natural history is book 7, on the singularities of the human species.

The curiosities of nature theme is said to develop along several lines: one, previously mentioned, evolves into prodigy books; the other, through Solinus's third-century epitome of Pliny's natural history (*Collectanea rerum memorabilium* or *Polyhistor*), Augustine's *City of God*, and Isidore's *Etymologiae*, lapses in the Dark Ages until the appearance of pseudo-Albert the Great's *De secretis mulierum* and Vincent de Beauvais's *Speculus majus*, and emerges in the modern prototype, the *Book of Secrets* (*Secrets de lhystoire naturelle*) of 1524, in which Pliny "remains the great purveyor of marvels."[27]

The anonymous author of the *Book of Secrets* seems less interested in understanding the things of this world, remarks one critic of monstrosity, than in renewing our contemplative wonder: "Whatever is common or ordinary seems negligible to him; whatever is exceptional or unique, on the contrary, captures his interest."[28] This exasperated search for the extraordinary and indifference to the regular or common are often presented as a legacy of Pliny's intellectual stance. Wrongly, I think. What Pliny seeks to impart to his reader is the tremendous variety of natural phenomena. The easiest way to do that is by choosing examples that are most unlike. This principle is elucidated in regard to nature's creative power: the creations that she accomplishes every day, almost every hour, are too numerous to be recorded on the scale of the individual, so Pliny jumps a category: "Let it suffice for the disclosure of her power to have included whole races of mankind among her marvels. From these we turn to a few admitted marvels in the case of the individual human being."[29] By discussing monstrous *races*, which require a greater creative effort on nature's part to produce, the compiler is relieved of the burden of enumerating all the individual anomalies.

But *anomaly* is an inappropriate term, because it is in the nature of nature to never repeat herself. Take human features: "Again though our physiognomy contains ten features or only a few more, to think that among all the thousands of human beings there exist no two countenances that are not distinct—a thing that no art could supply by counter-feit in so small a number of specimens!"[30] Methodologically speaking, then, it makes much more sense to concentrate on races, rather than on individuals, and on extreme cases rather than on minor variations: "Nor shall we now deal with manners and customs, which are beyond counting and almost as numerous as the groups of mankind."[31] And with this caveat Pliny launches into his famous, influential compilation of reports on monstrous races.

The most enduring trait to be sifted out of this tradition is, in fact, that of the monstrous races, for which Pliny's credulity, lack of rigor, and willingness to report any account, no matter how improbable, are often censured. Now the overwhelming impression formed by my own reading of book 7 of Pliny's *Natural History* is not of some superficial search after the anomalous, nor of an uncritical attitude toward sources.[32] Rather, the reader is afforded a rich bazaar of stories, facts, fables, anecdotes, and meditations to peruse; their truth-value is simply bracketed by deferring judgment and responsibility to the source from which they are taken. There is no question of Pliny's sorting out the fabulous from the real; this is simply not a pertinent criterion for his genre.

What stands out more than anything is rather an overwhelming sense of pessimism and weariness. Ingenious nature has created such a great quantity of monstrous races purely for its own *amusement* and for our *wonderment*.[33] Narcissistic nature is at times uncaring and cruel; her pro-digious variety is not a source of astonishment or terror so much as an amusing and bemusing pageant to behold. This relaxed, vaguely decadent atmosphere is evident in relation to the social status of hermaphrodites. At one time, Pliny observes, hermaphrodites were called androgynes and were considered prodigious, thus implying a sacred association, and mak-ing them a matter for the haruspices to interpret. Now, on the contrary, he notes with indignant irony, they are a source of pleasure or delight.[34]

The marvels of nature, and of human nature, must be seen, then, within the context of a "secular" philosophy. Pliny has no patience for philosophical or metaphysical speculations regarding the afterlife, or for theories on the immortality of the soul: "All men are in the same state from their last day onward as they were before their first day, and neither body nor mind possesses any sensation after death, any more than it did before birth," he states unequivocally.[35] One of the things that Pliny lists as unique to mankind, in a negative series of moral defects, is that he is, in fact, the only animal that is superstitious.

What we perhaps find difficult to reconcile is how a person who so uncompromisingly dismisses the fables of gods and concerns about an afterlife, and who disparages superstitions, would report so naively on the existence in Africa of families of sorcerers (*effascinantium*) who through their incantations kill newborn infants, dry up trees, and others who with a look alone are capable of bewitching and killing adults. Or on such monstrous races as the Astomi ("mouthless") or Apple-Smellers, who live on odors alone; the Gymnosophisti, who spend their days standing in fire and staring at the sun; the Dog-headed race, covered in fur and who bark instead of speaking, of which more than 120,000 existed at the time that his source (Megasthenes) reported on them; and so on.

John Friedman expresses our modern attitude perfectly: "The reader who has considered with some amazement the preceding catalog may wonder how the ancients . . . who traveled in these lands could have given credence to so many fabulous races of men." The question is particularly vexing because, as Friedman points out, direct personal observation had no effect on reducing the legends of monstrous races.[36] But this is a moot point: the chronicle of wonders genre does not arise from the same taxonomic motivation that fuels works of natural philosophy. Reality or fiction is not a pertinent factor.

The last brick in our survey of the foundations of monster literature is the work of Augustine. The *City of God* provided the most used source in regard to monsters for most Christian thinkers right through the seventeenth century. Three ideas are seminal. One, that since God created all things, and "he himself knows where and when any creature should be created," "even if a greater divergence from the norm should appear, he whose operations no one has the right to criticize knows what he is about."[37] Augustine also expresses this idea in terms of the Big Picture: "A picture may be beautiful when it has touches of black in appropriate places; in the same way the whole universe is beautiful, if one could see it as a whole."[38]

Deformity and monstrosity are simply in the eye of the beholder, then, since assuredly they fulfill some necessary function in the whole workings of the universe. Related to this view is Augustine's aesthetic of *concordia discors,* that "there is beauty in the composition of the world's history arising from the antithesis of contraries—a kind of eloquence in events, instead of in words."[39] Monsters are thus a rhetorically delightful event in the narrative of history.

This idea is carried through in Pseudo-Albert's *Secrets of Women:* according to the naturalists, he explains, monsters are made "to embellish and to decorate the universe. For they say that just as diverse colors existing together in a wall decorate this wall, so indeed, diverse monsters embellish this world."[40] Alexander of Hales (*Summa Theologica*) applies the same argument in his discussion of monstrous races: "So, just as beauty of language is achieved by a contrast of opposites . . . the beauty of the course of the world is built up by a kind of rhetoric, not of words, but of things, which employs the contrast of opposites."[41]

The concept of marvels of nature, or curiosities of nature, has little of the supernatural about it: monsters provoke neither horror nor terror, nor a need to perform ablutions: they are a particularly delightful rhetorical flourish on the part of God the Grammarian. Whereas Aristotle's grammarian erred in producing monsters, demonstrating nature's deficiency, Augustine's Divine Grammarian displays especial virtuosity, and in no way can he be viewed as an imperfectly skilled craftsman. Nature is, on the contrary, his instrument: "It is, in fact, God himself who has created all that is wonderful in this world, the great miracles and the minor marvels which I have mentioned; and he has included them all in that unique wonder, that miracle of miracles, the world itself."[42] The fact of the matter is, however contrary to nature or "unnatural" *monstra* might seem, God may speak to us in whichever way he pleases "with no difficulty to hinder him, no law of nature to debar him from so doing."[43]

This brings us to the second seminal idea in the Augustinian tradition. Nothing is contrary to nature: "We say, as a matter of course, that all portents are contrary to nature. But they are not. For how can an event be contrary to nature when it happens by the will of God, since the will of the great Creator assuredly *is* the nature of every created thing? A portent, therefore, does not occur contrary to nature, but contrary to what is known of nature."[44] This view sounds remarkably similar to seventeenth-century statements regarding the explainableness of natural laws: what seems miraculous only appears to be so to the ignorant; as soon as the workings of gunpowder are revealed, for example, it ceases to amaze us. Wonder is a companion of ignorance.

But "the enormous crop of marvels, which we call 'monsters', 'signs', 'portents', or 'prodigies' " still remain to be explained. To put it succinctly, they are both God's way of speaking to us and proof of his presence down here on earth: "These 'monsters', 'signs', 'portents', and 'prodigies', as they are called, ought to 'show' us, to 'point out' to us, to 'portend' and 'foretell', that God is to do what he prophesied that he would do." They must take place outside the realm of ordinary natural laws in order to retain their characterization of miracles, which is why they cannot provide material for divination, a faulty science in any case. Or if the forecasters do come up with accurate predictions, it can only be "under the influence of evil spirits [demons], whose aim it is to entangle in the toils of baneful superstition the minds of such human beings who deserve that kind of punishment," that is, to be afflicted with demon worship.[45]

This distinction will serve to delineate the realms of divine magic, natural magic, and demonic magic for writers such as William of Auvergne in the thirteenth century and many theoreticians of magic in the seventeenth century, whom we will be looking at in depth. Natural magic, as we shall see, works with and within the laws of nature. Demonic magic supersedes natural laws by intervention of supernatural yet material beings. Divine magic has no relation whatsoever to the understanding of man.

The third idea has to do with monstrous races as opposed to individual monstrosities. Augustine's discussion takes place in the context of the veracity of biblical history, that is, "whether we are to suppose that they [the monstrous races] descended from the sons of Noah, or rather from that one man from whom they themselves derived." He expresses great reservation at the pagan accounts ("If these are to be believed . . ."), which he refers to disparagingly as "portrayed in mosaic on the marine parade at Carthage, taken from books of 'curiosities', as we may call them." His policy is clear and unequivocal regarding the human status of monstrous races: "No faithful Christian should doubt that anyone who is born any-where as a man—that is a rational and mortal being—derives from that one first-created human being. And this is true, however extraordinary such a creature may appear to our senses in bodily shape, in color, or motion, or utterance, or in any natural endowment, or part, or quality." He further notes that the definition we make of man is crucial in deciding a creature's human status. "The definition is important," Augustine notes, "for if we did not know that monkeys, long-tailed apes and chim-panzees are not men but animals, those natural historians who plume themselves on their collection of curiosities might pass them off on us as races of men, and get away with such nonsense."[46]

What a Monster Is and Is Not

Given this intermingling between real instances of monsters, accounts of monsters, and histories of accounts of monsters, is there any utility to be derived from distinguishing between "fact" and "fiction" in this litera-ture? How are we to organize and sort out real monsters—babies born with birth defects, for example—from mythological monsters or poetic treatments of metamorphosis?

An answer becomes possible only if we are willing to approach mon-sters not as a thematic topic or as a psychological manifestation of some primal fear but rather as an "ideological cluster," as an entity constructed and represented within a social group. What constitutes a monster at any given time or place is a contextualized, localized set of characteristics, defined and accepted by the community. Monstrosity ranges freely from physical to moral qualities and back again, seemingly unconcerned by the different orders of reality implied in this conflation.

For example, Giambattista Marino, in one of his *Galeria* poems, likens Martin Luther to a "Hidra ferace / di mille avide teste [Fertile Hydra / of a thousand avid heads]."[47] But Luther himself makes great use of a famous monster, baptized the Mönchskalb (the monk-calf), whose image appeared in numerous publications, from his own religious propaganda pamphlets[48] to the scientific compendiums of Fortunio Liceti, Conradus Lycosthenes, and Ambroise Paré. From all accounts, this monster actually existed before it was turned into an emblem of Catholic depravity. Lycosthenes has it born in 1522 to a person named Stecher, while Paré tells us that Stecquer is in fact the name of the village in Saxony where it appeared, and Liceti insists that "a brute by the name of Stecher, was the father."[49]

These discrepancies would be inconsequential if it were not for the fact that Jurgis Baltrusaitis, a modern scholar of aberration, unproblematically lists this particular monster as an invented "bête fabuleuse," a "monstre raisonné" whose every detail is carefully constructed according to its allegorical meaning. The absence of any body hair indicates the seductive glimmer of hypocrisy; its fleshy "hood" illustrates obstinacy in heresy, and so forth.[50] Was the monk-calf a "real" monster, or was it invented by Luther for propagandistic purposes? Clearly the question is irrelevant when we consider the role it played during its time in the context of the culture that constituted it as a significant event.[51]

In this study, then, what is meant by a monster is not any particular *thing:* it is a category that becomes constituted in different ways according to different cultural and historical contexts. An actual "flesh and bones" creature displayed in a museum, a philosophical concept used to investigate the workings of nature, a rubric in a zoological taxonomy, a decorative figure in an emblem book—monsters are to be found in manuals of natural magic, in demonology treatises, in medical treatises, in discussions of the body politic, and in theories of wit. They flourish in museum collections of both *naturalia* and *artificialia;* they appear en masse in gardens, in gadgets, and as automatons; they perform in catoptric theaters and under microscope lenses. At times they seem strangely familiar, at others intensely disturbing. Let us begin, then, where the mythological parents of Western humanity first encountered a monster: in the garden.

TWO

Monstrous Matter

An interesting event took place in 1536 in the private garden of the
Rucellai family of Florence. Now many interesting things happened in
the Orti Oricellai or Oricellari, as it was called. Some of the events were
ritualistic and hedonistic—sumptuous festivities and the performance of
the first tragedy in Italian, "Rosmunda," for example[1]—while others displayed the seriousness of philosophical gatherings. In the times of Bernardo Rucellai (1448–1514) the Platonic Academy instituted by Marsilio
Ficino and Lorenzo Il Magnifico continued to meet there. Discussions on
literary and philosophical topics among the greatest humanists of the age
shifted into more contemporary concerns regarding the ideal form of
government. This sort of discussion later focused upon the contributions of another illustrious member of the Orti Oricellai group: Niccolò
Machiavelli.[2]

Bernardo's opposition to both Medici tyranny and populist democracy
eventually led him into exile; the meetings were suspended, and on his
return he retreated to his garden and retired from public life. No doubt,
wandering among the rich collection of exotic plants and classical busts
and statues that peopled the vast garden on Via Scala between the second
and third circles of the Florentine walls, the aging oligarch was able to
renew himself in contemplative *otium* (leisure) as he reflected on the vi-
cissitudes of political fortune. The meetings were resumed, clandestinely:
angry young men met there to vent their criticisms of the regime.

But after the restoration of the Medici in 1512, public meetings recon-
vened, this time presided over by Bernardo's nephew Cosimo and his sons
Palla and Giovanni, active political and literary figures. The garden, then,
was a place for the "Ottimati" of Florence[3] to gather and talk about issues
of philosophical, historical, and political import. The Orti Oricellai school,
avidly welcoming Machiavelli back from exile, became synonymous with

the rise of modern political history. It also lent its name to a set of seditious conspirators whose 1522 plot to assassinate Cardinal Giulio led to Jacopo da Diacceto's beheading, and to the exile of many of its members. With this series of events, the meetings ceased.

Later, in 1578, the Orti Oricellai was the scene of a titillating, extravagant "happening" designed to alleviate the tedium of the reclusive Grand Duke Francesco (1541–87) and his aristocratic guests, an event immortalized by Malaspini in one of his *Novella*.[4] The decadent, culturally eclectic atmosphere of the Saturnine Grand Duke, lover of mechanical wizardry, enthusiastic alchemist, and patron of museums, is vividly evoked in Malaspini's prose.

The guests were greeted by a bizarrely clad necromancer. After drawing a magic circle in the ground with a knife, the wizard summoned four devils with Dantesque nonsense names and tossed a nauseating sulphur mixture onto the fire. The "glow of charcoal afire in the oils" provided, in fact, the only eery light available at 1:30 in the morning. The grand duke and his guests, although tremendously amused by the necromancer's antics, were almost overcome by the stench. The tortures (reminiscent of similar fêtes staged by de Sade's infamous Juliette) had just begun, though: the guests then had their hearing assaulted by "infinite voices and laments, strange yowlings, gnashing of teeth, hands clapping, shaking of iron chains, cries, sighs, and infinite fireworks that exploded from all sides, issuing forth from many holes dug with marvelous artistry."

This marvelous combination of terror and delight is typical of the age. In this particular *festa* it reached an apogee. The *negromante* then set off a mine and on his signal the earth literally dropped away to deposit the grand duke and his guests in an even more Dantesque underground re-creation of hell. The guests were so upset, writes Malaspini, that they didn't know whether they were dead or alive. But, as Eugenio Battisti notes in his analysis of the *beffa*, or practical joke: "after the Christian hell, an Islamic paradise."[5] A group of beautiful young girls appeared to lead the company out of the stench toward the sweetly perfumed terrace of the garden, where angelic voices piped out a marvelous concert of madrigals. Under their golden mantles, studded with pearls, diamonds, rubies, sapphires, and emeralds, perfumed from head to toe, the girls were entirely nude. The guests feasted their eyes on the "most graceful young women" and as a consolation, after "they heard a huge roar and thundering, and with a mighty impulse the devils were expelled from the garden," the party went on to feast their bellies on a sumptuous dinner.

This elaborate ritualistic event, staged in the garden, evokes other atavistic initiatory rites: the members of the party are systematically subjected to sights, sounds, and odors that confuse their senses, to the point that they are unsure if they are alive or dead. Battisti sees this as a way of removing any rationality from the spectators, of inducing a dreamlike state of mind, so that reality and artifice begin to intermingle: what is play and deceptive illusion, what is true and natural?

The garden clearly functioned in this case as a liminal space, where transgression was licit and worldly values were turned upside down. The monstrous statues of Bomarzo, Pratolino, and Castello were fitting accoutrements for this dreamlike pilgrimage, this "initiatory process, from Inferno to Eden" so often planned and staged in late Renaissance and Baroque gardens.[6]

Benedetto Varchi and the Double-bodied Monster

The event I would like to focus on was a meeting of a different sort entirely, neither literary nor political nor festive in nature. What I am referring to is the day in 1536 that Palla Rucellai, Alessandro da Ripa, Francesco da Monte Varchi, "& a few other excellent Physicians, & Painters" gathered in the Rucellai gardens to dissect a monster. The event and the monster in question are described twelve years later by Benedetto Varchi (1502–64), a one-time opponent of Cosimo I, later granted a stipend by the grand duke, named official historian, and set up in the Medicean villa La Topaia. Varchi remembers the incident in a very scholarly disquisition of some fifty printed pages presented publicly to the Florentine Academy on two successive Sundays in July of 1548. It is called "On the Generation of Monsters & Whether They Are Intended by Nature or Not."[7]

The story is told in chapter 1 in the context of a discussion on what monsters are, their formation, and how they are produced. Varchi begins by affirming the reality of monsters: "We confirm that it is absolutely true both in Animals as in men that monstrous births occur, which either have too many or lack ordinary members, both external ones and internal ones, or their members are transposed or damaged."[8] Leaving aside "those that you hear of all the time in the histories," Varchi—a man who bases his theorizing entirely on Aristotle's *Physics* and *De generatione animalibus,* and a firm believer in seeing with his own eyes—starts out by devaluing textual evidence and by bringing the burden of proof much closer to home: "But must we recount what others write? Not only have many monsters been seen, both long ago & in our own times, not only in Italy (like the one in Ravenna) but even in the Florentine Realm, & in Florence itself."[9]

He then appeals to his colleagues, members of the audience gathered that Sunday to hear the eminent scholar: "There are so many people in this place who remember having seen that Monster that was born at the gates of Prato, some twelve years ago. . . ."[10] The description and the story are offered, then, as proof of Varchi's personal authority, and as specific and objective evidence of his theory of monstrous generation.

> [They were two females [he explains] joined & stuck together, one toward the other in such a way that half the chest of one along with that of the other made up a single chest, & thus they formed two chests, one joining up with the other; their backs were not shared, but each had its own: it had its head turned directly toward one of the two chests, & on the other side, in the place of the face it had two ears that were joined one to the other, & they touched: the face was very beautiful: blue eyes: it had upper teeth, & the lowers of one girl were extremely white, softer than bone, & harder than gristle, as big as a man's; she was very well proportioned; the other girl, from mid-back down, was twisted, & especially her legs, which were very short in comparison with the other girl's; she had a certain purplish skin that covered her back, & it came forward all the way to her private parts, adhering to the pubic region; both of their arms, & hands were very beautiful & well proportioned & appeared to be, like all the other members, about ten or twelve years old, even though the Monster was young. The girls were separated at the navel, which was the sole source of nourishment for both.]

> Erano due femmine congiunte & appiccate insieme l'una verso l'altra di maniera, che mezzo il petto dell'una insieme con quello dell'altra, facevano un petto solo, & cosi formavano due petti, l'uno rincontro l'altro, le schene non erano comuni, ma ciascuna haveva le sue di per se: haveva la testa volta al diritto dell'uno de' duoi petti, & dell'altro lato in luogo di volto haveva due orecchii, che si congiugnevano l'uno contra l'altro, & si toccavano: il viso era assai bello: gli occhi azzurricci: haveva i denti di sopra, & di sotto bianchissimi piu teneri, che l'osso, & piu duri, che il tenerume, grandi come d'huomo una delle quali era molto bene proporzionata, l'altra dal mezzo della schiena in giu era stroppiata, & specialmente le gambe, le quali erano molto corte a comparazione dell'altra haveva una certa pelle pagonazziccia, che la copriva di dietro, & le veniva dinanzi infino alla natura, appiccandosi al pettignone; le braccia, & le mani d'entrambe erano bellissime, & ben proporzionate, & mostravano, come tutte l'altre membra di dieci, ò di dodici anni, ancora, che'l Mostro fusse piccolo. La separazione di dette fanciulle era nel bellico, il quale solo serviva al comune nutrimento d'amendue.[11]

It is difficult not to be struck, in this prose passage of a single sentence, by the insistence on parallel structures of unequal elements, as if to emphasize the marvelous incongruity of "same yet different" and "one yet two" that the monstrous twins exhibit, as if the monster's body alone were able to contradict the principle of noncontradiction (i.e., a term cannot be itself and other). "They were two females" with "two ears," "each with its own spine for itself," yet they are also one being "joined and attached" with "a single chest"; yet the one chest "thus formed two chests." "The one" and "the other" echo off each other in a ping-pong rhythm throughout the passage: "l'una verso l'altra," "una insieme con quello dell'altra," "l'uno contra l'altro," "l'una rincontro l'altra," and so forth.

The twins share a common source of nourishment through the navel, yet the *bellico* is also their point of separation. They are similar, yet they are different: one is perfectly formed, the other lacks a head and is deformed from the mid-back down. They are both extremely beautiful, yet unspeakably ugly. They are very well proportioned in parts, their teeth are "extremely white," the eyes blue; "both of their hands are extremely beautiful," yet one is covered in a purplish skin. They are both little and big, young and old. Incredibly, their hands and other limbs "had the appearance of ten or twelve years old" even though the "Monster was young." "Above" and "below," "more tender" and "more hard," Varchi's description is a densely packed set of attributes that compares and contrasts while ranging freely over the extension of the visible body.

A prose of comparison, a prose of astonishing perplexity: the monster challenges Varchi both conceptually and stylistically, stretching the use of parallel structure to the limits of intelligibility. When we have finished reading this passage, difference and identity, one and other, two and one, have been exchanged so often, and played against each other so incessantly, that they begin to resemble each other. It is a matter of indifference which one is which: we are more entertained by the fascinating conceit (or witty figure of speech) that the monster offers our imagination than we are in forming an accurate picture of it in our minds.

Galileo is usually accredited with the development of Italian scientific prose, some fifty to seventy years later. Let us say, then, that Varchi's contribution belongs to a highly creative and individualistic period of stylistic experimentation in which the semantic resonance of words, by complicating and problematizing the task of representing the referent, by adding poetic and conceptually unrelated interferences, is not viewed as obstructing communication.[12]

Certainly such a detailed description of the exterior aspects of the monster, born twelve years earlier, also bespeaks careful notes taken at the time, a passionate interest in detail for the sake of accuracy, and a dispassionate eye that views and seizes the corporeal, material appearance of the creature as specimen and not only as *concetto* or trope. The fact that the great painter Bronzino was not only present in the group but also applied himself to producing an "illustrious portrait"[13] of the monster is not purely an index of the Mannerist penchant for the grotesque.[14] Bronzino's contribution, from one member of an interdisciplinary group composed of "most excellent" physicians, painters, and natural philosophers, must be seen as "documentation," as scientific illustration.[15]

But the group did much more than describe the monster they had in front of them. Having retrieved it from some woman's bed chamber, presumably stillborn,[16] they brought it to the Rucellai gardens to dissect it. Says Varchi after this minute description of its exterior: "Fecesi sparare nell'horto di Palla Rucellai." They cut it in half and examined its insides: "They found two hearts, two livers, & two lungs, & finally everything was doubled, just as for two bodies, but the windpipes, which began at the hearts, joined up near the entrance to the throat & became one: Inside the body there were no divisions, but the ribs of one stuck to the ribs of the other all the way to the pit of the stomach, & from there down it served both of their lower backs."[17]

We can note briefly the reprise of the "two yet one" theme, but the lexicon has shifted, using more specialized anatomical terms ("windpipes," "fontanelle," "ribs," "pit of the stomach"). It would seem that the anatomical interior, a far rarer sight, offering less common objects to view, is more easily tamed and appropriated by a lexicon that has been formed to describe, precisely, these particular objects. Varchi's tone takes on the assurance of a jargon.

The academician finishes this story with a final irrefutable bit of evidence based on a wider application of the logic of comparison: because of "these & many other similar Monsters & different ones, like those that you see in the loggia of the Scala Hospital, we philosophically believe that there have been, & can be monsters."[18] Thus, this double-bodied monster serves as a token of a type against which other particular monsters can be compared for differences and similarities. The hospital wall of La Scala was apparently decorated with votivelike figures of unusual fetuses that had passed through its corridors, very much like those mentioned by Augustine as portrayed in mosaic along the marine parade in Carthage.[19] These specimens provide the proof that "siano stati"; tokens and types, comparisons and contrasts offer the proof that, "filosoficamente," they *can* exist.

What interests me about this incident is not only how the event is transformed into a textual occasion, and how that textual event transforms the monster into a scientific object; I am also fascinated by the location, and by the significance of the act of dissection. What does it mean to bring the monster into the garden? Why dissect it there, instead of, for instance, in the hospital or in a field? What is implied by the act of dissection and description? How does description differ from other possible treatments of monstrous births, like burning the infant as an abomination, or allegorizing it, viewing it as an emblem of moral vices and virtues, or employing it for prognostication or propaganda purposes? All these were possible courses of action available at the time the incident took place and when Varchi chose to write about it.

Bringing the Monster into the Garden

By bringing the monster into the garden—ancient archetype of Golden Age and Edenic visions[20]—the Oricellai group signaled the split and deviation of two distinct types of spaces in which botanical ordering took place. They also opened the possibility for another.

One type of garden traces its genealogy back to the medieval *hortus conclusus,* inside of which gardeners and herbalists produced their plants for nutritional and pharmaceutical uses. This primarily utilitarian type was embodied in the enclosed gardens of monasteries, which cultivated grapes, olives, vegetables, fruit trees, and so on for their kitchens; and in those of herbalists, who also reserved space for experimental agricultural techniques and for precious plants, cultivated for their magical or medicinal properties. These are the "secret gardens," growing within the enclosure of walls, favored by the structure of the medieval city.

Only in the fourteenth and fifteenth centuries did gardeners begin including plants for the beauty of their flowers, their colors, their rarity. Gradually the garden lost its utilitarian aspect, later functioning as an instrument for organizing knowledge. With the first chairs of *lectura simplicium* in the university, the hospital herbalist was promoted to the status of learned botanist. What was previously a place of production for medicinal plants became transformed into a device for the production of knowledge, an early text for the dawning of botanical science. This evolved into the pedagogical tool of the *orto universitario* or *jardin des plantes.*

The other type of garden—*l'Orto del Principe*—was primarily hedonistic or aulic in nature, a necessary adjunct for "i regi virtuosi" or princely experimentalists, as Agostino Del Riccio (1541–98) describes them in his treatise "The Kingly Garden." The garden should provide "tranquillity and relaxing silence," Del Riccio proposes, a place "to relax by reading some entertaining poetry." It matters not whether the garden is "made with an artful hand" or left in its natural wild state, it is a place one frequents to renew oneself, to take a break from the company of men.

Other essential components of the princely garden include a *serraglio*, a living collection of animals. When princes tire of reading poetry, watching such exotic possessions as lions, leopards, giraffes, or ostriches, they might take equally great delight in listening to song birds or gaming with hunting birds. And if in their "honesta taciturnità," the princes begin to crave the company of others they might gather there "to joke together with entertaining and loving games that appear to gibe each other yet really no harm is done to anyone." Such was the *beffa* arranged by Grand Duke Francesco's engineers for his amusement that was described above. More often, though, the prince himself constructed automatic sculptures and fountains that would squirt water at unsuspecting and hapless spectators.

But the main function of the garden was to provide a place apart for the prince, away from distressing contact with his courtiers and subjects, described in rather unflattering terms by Del Riccio: "O what pleasure these most pleasant and delightful places provide, which are granted to prudent princes that are plagued by so many offences and displeasures in ruling so many, often insolent, disobedient, aggressive, lying, thieving, homicidal peoples."[21]

This hedonistic luxury garden attached to a villa sometimes took on a more scientific cast, depending on the tastes of the prince. Increasingly exotic species were imported and arranged to produce a *theatrum naturae*, a highly ordered space analogous and often ancillary to a princely *studiolo* or *museo*, designed to represent all species of nature in a single organized space. A theater of nature was like a Noah's arc of the intellect, aiming to capture at least one example of every plant and animal in its panoptic vision. It also offered the possibility to tame nature's productions by imposing a classificatory scheme that made relations between parts and the whole intelligible and significant. These sorts of museum gardens merged the classificatory functionality of the botanical garden with the escapist possibilities that the secret garden of the villa offered the princely collector.

As an Edenic otherworldly space for the Platonic Academy's contemplation, the Orti Oricellai provided splendid isolation for the Florentine court. It exactly fits Marcello Fagiolo's ideal description of the princely garden: "Immersed in greenery, it is both an invocation of the lost paradise and an ideologization and prefiguration of a well-ordered society." The Orti Oricellai also followed the transformation Fagiolo describes "into the idea of a citadel of leisure [*otium*]." Bernardo, withdrawn from public life, retires to his garden, peopled now with "mysterious statues" of gods and heroes. Governing members are figured as ancients or vice versa; statues of divinities are represented with the features of the governing class—a modern head, for example, on a classical bust.[22]

This political allegory was truncated, literally, by the decapitation of Jacopo da Diacceto in 1522. The later Orti Oricellai of Palla's time thus floats indefinitely between a mythical space of communion with Edenic–Golden Age pretensions and an ordered political model, enacted in the arrangement of the classical statuary that peopled the garden and in the group's systematic discussions around Machiavelli's dialogue with the ancients in his *Discourses on Livy*.

The dissection of the double-bodied monster thus signaled the possibility for another, new function for the garden and for the oligarchic gentlemen who found themselves cut off from public service and political power after the restoration of the Medici: it transformed the Orti into an anatomical theater, into an experimental laboratory. In this context, Bronzino's portrait of the monstrous infant loses its superfluous, decadent grotesqueness and takes on the utilitarian function of documentation. The meeting itself takes on the lineaments of an Accademia del Cimento, a Royal Society, and *otium*, or idle curiosity, is transformed into a respectable experimental pursuit.

This moment of "pure science" is, of course, no final term in an evolutionary process; it is simply one moment among others. The garden as a liminal space for sensual, initiatory rites, as a mythological and transgressive locus, returned in the pagan "performance art" staged for the Grand Duke Francesco. As we shall see, scientific and technical experimentation was often, if not always, associated with sacred rituals, religious institutions, or the pursuit of pleasure and amusement, rather than with utilitarian production. The history of garden automatons is a fertile example of this line of thought. But for now let us return to that bloodied, mutilated corpse, so lovingly and minutely described by Benedetto Varchi.

The Nature of Description

As I have said, recuperation, dissection, and description of the monstrous infant constituted an anomalous reaction given the possible courses of action available at the time. Far more often the fate of such a monster was quite different: burning, drowning, suffocation, if it lived that long. Varchi himself recounts an incident contemporary with his talk: "that Monster, that was born in the year 1543, in Avignon . . . which . . . had a human head from the ears out, together with the neck, arms, & hands of a dog, & also its virile member." Varchi brings up this case in order to debate whether humans coupling with animals can produce hybrid off-spring. His conclusion, along with Aristotle, is negative, but evidently not everyone subscribed to the same natural philosophy. In the face of the physical evidence ("All the canine members were covered in long, black fur, like the dog's, which the woman who gave birth to the monster later testified to having lain with"), the infant's resemblance to a dog and its efforts to speak that ended in barklike sounds, "it was brought from Avignon to Marseilles to the Most Christian King Frances, who on the last day of July had the woman & the dog burned together."[23]

Let us call this reaction pious terror: the fact that the woman confessed to having had intercourse with the dog sounds very similar to contemporary confessions of witchcraft, extracted under torture after accusations had already been formulated. Burning was a sentence for those convicted of communicating with diabolical beings. We are justified, then, in supposing that this monstrous infant was viewed as a product of sacred forces. The "purely scientific" attitude, which turns the monster into a specimen, on the contrary, evacuates it of any terror or metaphysical significance. Burning destroys the physical object, obliterates the evidence, purifies the arena, and expiates the sin. Curiosity creates an object, turns it into an illustration, places it in a rhetorical context, and makes it part of the theater of nature in which every object has its place in a meaningfully organized cosmos, at once representable and legible to the spectator of knowledge.

By being described, the double-bodied monster was in a way transferred from the category of *naturalia* into that of *artificialia*. Placed in a glass jar, arranged on a shelf in the museum, it could be collocated in the same category as Bronzino's portrait, such as the one he completed of "Morgan the dwarf viewed from behind," which would be most appropriately ordered alongside Giambologna's fountain-piece "Morgan the dwarf riding a dragon."[24] Engravings of Vincent Levin's museum dating from 1706, in fact, clearly show double-bodied and other deformed fetuses peeking out of big glass containers, and arranged in rows on shelves, covering the walls (see figure 1). Above the shelves we see indigenous artifacts from the New World; below, paintings of birds, carved ivory balls, agricultural instruments, and jewelry.[25]

Description not only describes, it also creates, orders, sets the object in a context of rhetorical meaning and institutional forces. And finally, it also mutilates the specimen. Purely as a "gedanken" experiment, imagine this same monster born during the Roman Republic. Let us unfocus our eyes a little, blur the image, and step back some distance. An animal of some kind is brought into a field; a group of eminent, highly learned men appears and gathers around. They examine the creature, take careful

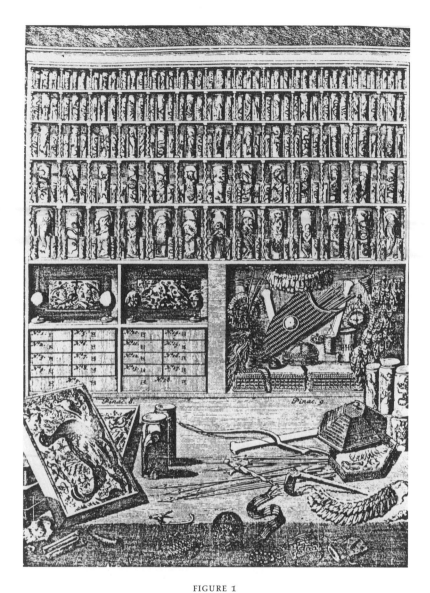

FIGURE 1

Glass jars with monstrous fetuses displayed along with ethnographic
curiosities, jewelry and other *artificialia* in the Levin Museum.
From V. Levin, *Wondertooneel der Nature*, 1706.
(Reproduced from Adalgisa Lugli's *Naturalia et Mirabilia*
[Paris: Éditions Adam Biro, 1998], p. 66)

notes, discuss, and then retire to make their findings public: the creature lacks a head; the king will leave no successor to the throne. Or, they dissect the creature, cut it in half, and carefully, minutely, examine the state and disposition of the internal organs. They record their observations and announce their findings: the liver is glassy, a propitious sign; the war should begin.

Now return to our Rucellai gardens, where the top members of the professional classes gather to examine the monster: run through Varchi's account from a similar distance, producing a sort of pantomime effect, and then read the description again. Perhaps some would draw great distinctions between the actions of the ancient haruspices and those of the Florentine Academy. No doubt, they stem from different worldviews. But one thing is clear: in both cases the specimen is created as an object of significance; it is mutilated, described, and placed in a rhetorical and institutional system of knowledge.

Philosophy and the Material Origins of Monsters

Varchi's lesson is not only a series of descriptions. Its primary interest is a philosophical question: whether monsters are intended by nature or not. Whenever monsters are subtracted from the metaphysical realm, whenever their status as divine messengers is removed, either by denying their value as prognostic signs or by debunking the idea that they are punishments from God (arguments that continually surface at various times and places, from Lucretius to Augustine to Liceti), then the presence of monsters in the natural order presents a genuine puzzle to thinking men.[26] If monsters are not *monstra*, in the Etruscan sense, and if they are not meant to signal the wrath of a deity, in the Christian sense, then what possible significance can they have? Or if they have no meaning at all, then why do they exist?

An answer hinges on the production of natural beings. To understand their presence, the actual existence of these creatures, it seems logical to begin with how they come into being in the first place. From their effect (their material existence), looking back to their causes (their generation), one hopes to understand their purpose (their reason for being). Origins are thus seen to hold the key to explaining the presence and purpose of monsters in the sublunary world.

So the question of whether nature produces monsters intentionally or randomly, or whether she simply "slips up" occasionally is a profound one that exercised the intellects of some of the greatest natural philosophers of Western culture, beginning in the first recorded writings with Empedocles, addressed clinically by the Hippocratic writings, and elaborated at length as both a philosophical and physiological problem by Aristotle. The idea of nature committing errors, sinning, being deficient in some way, or acting randomly, holds far-reaching implications for human beings, whether the errors are conceived as part of the natural order or apart from the natural order. Exploring deviations from the norm is thus not only a way of better distinguishing the regularities of (posited) natural laws. Deviation is also intrinsically interesting, in itself, for what it shows about the purpose of nature's creativity, the workings and the character of nature in general, as well as humankind's status and role in the phenomenal world.

Embryological development of the individual animal or plant is projected onto the origin of the species, and vice versa: the earth is often figured as matrix or womb; the origin of birth defects is sought in the original appearance of diverse and monstrous species at the dawn of time. What theoretical treatise on monsters does not at some point grapple with the enigmatic fragments of Empedocles,[27] which suggest a bizarre creation theory (based on the perpetual dialectical struggle between Love and Strife) during the first reign of Love over the earth after the chaotic dominion of Strife?

> Here many neckless heads sprang up.
> Naked arms strayed about, devoid of shoulders,
> And eyes wandered alone, begging for foreheads . . .
> But when god mingled more with god
> these things came together as each happened,
> and many others in addition to these were continuously born.
> Many grew double-headed, double-chested—
> man-faced oxen arose, and again
> ox-headed men—creatures mixed partly from male
> partly from female form, fitted with dark limbs.[28]

These passages were rarely interpreted allegorically, although there were some commentators who suggested that they should be. The actual existence of hybrid-seeming creatures, along with reports of monstrous races, ensured that the pronouncements of the ancient shaman-philosopher could not merely be relegated to mythopoetics.

Lucretius provided another standard image of the earth as moist womb, from which animals and plants were generated "with the aid of showers and the sun's genial warmth":

> There was a great superfluity of heat and moisture in the soil. So, wherever a suitable spot occurred, there grew up wombs, clinging to the earth by roots. . . . Then nature directed towards that spot the

pores of the earth, making it open its veins and exude a juice resembling milk, just as nowadays every female when she has given birth is filled with sweet milk. . . . Then, because there must be an end to such parturition, the earth ceased to bear, like a woman worn out with age. . . . In those days the earth attempted also to produce a host of monsters, grotesque in build and aspect—hermaphrodites, halfway between the sexes yet cut off from either, creatures bereft of feet or dispossessed of hands, dumb, mouthless brutes, or eyeless and blind, or disabled by the adhesion of their limbs to the trunk. . . . These and other such monstrous and misshapen births were created. But all in vain. Nature debarred them from increase.

Lucretius denies the implications of Empedocles' vision of stray, single limbs seeking partners with which to produce heterogenous hybrids: If anyone pretends that such monsters as a Centaur or a Scylla or a Chimera exists "he is welcome to trot out a string of fairy tales of the same stamp." The facts prove otherwise: "The fact that there were abundant seeds of things in the earth at that time when it first gave birth to living creatures is no indication that beasts could have been created of intermingled shapes with limbs compounded from different species . . . each species develops according to its own kind, and they all guard their specific characters in obedience to the laws of nature."[29]

I have provided these quotations from antiquity in order to evoke the venerable and continuous philosophical tradition that thinkers like Benedetto Varchi assumed and interacted with. Since monsters, as we have seen, pose wide-ranging interdisciplinary problems, and since philosophy and inquiry into natural processes were only divided into separate disciplines during the seventeenth century, our investigation into "scientific" literature of early modern Italy will also perforce be an investigation into several thorny, perennial, and dense philosophical issues. In order to understand what sacred monsters turned into, or where they went to, once they were displaced from their status as portents or prodigies, we must look into the Aristotelian formulation of the categories of matter and form and, more importantly, how those categories were interpreted and adapted to the seventeenth-century worldview.

The Description of Nature

Varchi's approach to monsters, I have said, can be characterized as descriptive. I would further suggest that the act of description constitutes what we may call the "scientific" attitude, bearing in mind both the historicized, contextual understanding of this term, as well as its problematized use in contrast to "religious" and "magical" modes of thought.[30] We must also keep in mind that description is not simply a passive act in regard to the object being described: its institutional and interventionist character was brought to light in the Orti Rucellai incident. Still, the principal aim of a descriptive enterprise is to classify and create taxonomies. This is what histories of teratology generally call scientific treatments of monsters, and this is what, in its "purest" form, stands at the basis of comparative embryology and abnormal pathology.

The scientific attitude shown in the descriptive enterprise, the desire to describe a monster in this way, implies a concomitant evacuation of pious terror. This possible interpretation of monsters directly results from the definition of natural beings and their origins adopted in the philosophical context in which the monster appears. To be more specific: the "scientific" approach of Aristotelian thinkers starts out with the assumption that *being (soul, spirit, or anima) is a product of the arrangement or organization of matter.* This assumption is what leads to an interest in describing the exact arrangement of limbs and organs of the anomalous being. Other formulations of the relation of spirit to form and matter give rise to different reactions, which we will examine later.

A patient reading of Benedetto Varchi's lesson will not only make this evident, it will also set up the parameters for the discussion that follows. Varchi begins his *Proemio* by summarizing the Aristotelian causes, making it clear that the Final Cause is "la cagione delle cagioni," the cause of causes, and that "this ultimate final cause is so necessary that all the effects that it does not treat, even if it has all the other three causes, Efficient, Material, & Formal, for this reason cannot be called truly natural, since they are not intended, that is to say, ordered & desired by Nature, but they are born of fortune & chance, as produced temerariously & haphazardly, beyond the will & intention of the producer."[31] Neither artificial nor natural beings ("all things, including those produced by art, that are generated by Nature") can be products of chance or fortune. By definition they must have a final cause, a reason for being.

Now monsters would thus seem to contradict this definition. Like thunder, earthquakes, and other such prodigies, "being filthy & wicked things" they are apparently "errors & sins on the part of those that make them."[32] However, this is impossible since "we cannot think, nor should we, that they are either intended or desired, neither by GOD, who cannot err, nor by Nature, who never sins."[33] But this is challenged by the fact that nothing can happen in this world without God's knowledge and will; furthermore, Nature not only generates monsters, she also nourishes them and preserves them. Monsters undeniably do exist. Hence we are forced into the conclusion that monsters "are produced by Fortune & by chance."

This unacceptable view has been adopted and defended by countless "interpreters of Nature, both ancient and modern, & so many Greek, Arab, and Latin authors." In fact, what makes this aporia so worthy of investigation for Varchi is precisely its timeworn, venerable character, attested to by the company that it has kept over the ages. Varchi, like Aristotle (but for different ideological motives), is primarily interested in proving that "Nature does not operate by chance, as it seemed that DE-MOCRITUS, Empedocles, and several other ancient Philosophers would have it." In order to refute the argument that monsters display random operations in nature, Varchi-Aristotle turns the syllogism around.

Assume that monsters are errors and sins of nature. Now, if Nature "did not operate according to some end, but by chance, monsters would not be able to be called sins or errors" because the erroneous can only exist when there is some intention to produce the correct. Errors imply something done "in vain" (to use Aristotle's terminology). Where there is no intention there can be nothing done in vain. Hence Nature operates according to intention. The same argument holds true of art, "which without doubt operates for some end, and even if it errs at times like a Grammarian, who does not always write well, or speak correctly, and a Physician sometimes gives a medication that doesn't work, or works opposite to the intention of the Physician."[34]

Let us provisionally admit, then, that monsters are indeed errors of nature. This has the added attractiveness of debunking any possible notion that "contingent future things" can be known through divination, without however allowing for a meaningless world of phenomena. Varchi reports that "several other authors say that Monsters are produced to signify & announce future things, alleging events that one reads to have followed after such portents & prodigies in all the histories, & the customs of the Romans, who had them either burned, or thrown into the sea, or carried to some deserted Island & abandoned in order to placate the wrath of the Gods, & to escape the impending danger by order & commandment of the Haruspices."[35] His response to this befits an Aristotelian theoretician on monstrosity. Varchi dismisses the possibility that monsters are sacred messages or portents from God: "This was a superstition, as we read of many more in that religion & others."

Still, a problem remains: "But we must understand before we proceed further that this name Nature . . . signifies beyond universal Nature, that is to say, GOD, particular Nature; & this is divided into two, into form, which is the agent, & into matter, which is passive. And I doubt if Monsters are defects of Nature, whether we understand by this the universal or particular kind. And whether, if particular, it is a defect of the form, or the matter, or both together."[36] To summarize and paraphrase: the concept of "nature" has been used universally to mean all that exists, synonymous with God and the created world. It has also been used particularly, as when we inquire into "the nature of things," or say that something is what it is "by nature." The particularist sense also breaks down into two further categories. As Aristotle showed, when we ask what the "nature" of a thing is we sometimes answer by referring to its matter—the passive female element—"the immediate material substratum of things"; and sometimes by referring to its form—the active male element—"which is specified in the definition of the thing." "As an indication of this," Aristotle explains, "if you planted a bed and the rotting wood acquired the power of sending up a shoot, it would not be a bed that would come up, but *wood*—which shows that the arrangement in accordance with the rules of the art is merely an incidental attribute, whereas the real nature is the other, which further, persists continuously through the process of making."

Yet on the other hand, "We should not say . . . that there is anything artistic about a thing, if it is a bed only potentially, not yet having the form of bed; nor should we call it a work of art. . . . Thus in the second sense of 'nature' it would be the shape or form (not separable except in statement) of things."[37] Which definition of nature are we referring to when we say that monsters are errors of "nature"? The universal or the particular, and if the particular, is it the form or the matter that is deficient?

Varchi concludes (and here the argument takes a somewhat confusing turn since he seems at first to contradict his provisional premise that monsters are indeed errors of an intentional Nature) that monsters are not intended by Nature in *either* sense: "We can finally conclude, that monsters, being errors & defects, are not intended, either by universal or particular Nature, which cannot err, but by fortune & chance."[38] They are not errors of nature, because the deficiency does not lie in nature's intention, in her final cause, but rather in the *formal* and *material* causes.

The academician is completely convinced of the truth of this assertion: monsters are indeed products of chance and fortune, but in a different way than the atomists and Epicureans would understand it. Chance is removed from the universal realm and confined purely to the incidental, material order. Nature does her best; in no way should she be made to bear the brunt for occasional lapses: "She does not do it through her own fault or defect, rather she is obstructed by others."[39]

And who are these "altri" that hinder and impede nature from carrying out her purpose? The formal male agent of semen, acting on the passive female matter of the menses. "In this same manner, can we say of Nature, who never errs on her own, because if the semen is indisposed, & she does that which should be pardoned, & if the menses are not enough, or such, as much, & as what is required, whatever it may be, how can we blame Nature?"[40]

Varchi offers an analogy. If a sculptor wants to form a statue and is unable to do so because he doesn't have the material, or because the material is either "too hard or too tender, that it is unable to be made into a statue," we would not be justified in saying that the sculptor erred. Only "when he does not accomplish his end through ignorance of his art . . . could it be properly said that he erred." The point—and the solution to the aporia—is that when Aristotle called monsters errors and sins, "this does not imply other than lack of order, & lack of accomplishment of an end, & in sum, a deviation from its usual course and ordinary habit, which nevertheless does not arise due to the fault or defect of Nature."[41] When it comes to monsters, nature is like a tailor who either lacks a sufficient amount of cloth, or is presented with poor-quality material, and consequently "botched a garment"; "truly Nature lacks, that is to say her effect is defective, & she has too much or too little or something other than what she should & is customary to have."

We can see how clearly this formulation underlies and structures the description of the Rucellai monster. Its fully formed teeth that are "more tender than bone and more hard than gristle" echo the sculptor's material that is "either too hard or too tender [ò tanto dura, ò tanta tenera]"; the attention paid to two organs versus single ones views them as an excess for one person, but a lack for two, just like nature, which "has too many or too few or other than that which it should, & is customary to have [ha ò piu, ò meno, ò altramente di quello, che doverrebbe, & consueto d'havere]"; the exact arrangement of the interior organs bespeaks a "lack of order [mancamento d'ordine]"; the stress on well-formed limbs versus deformed limbs points to nature's efforts to do the best she could with the defective material at hand; the girl who is "stroppiata," or botched from her midriff down, repeats the motif of the "veste stroppiata."

In short, all these factors are read as indices—not in order to interpret the significance of the monster as the superstitious haruspices would have done, only to then command that the creature be burned, drowned, or abandoned on a deserted island, but rather to divine its nature, its purpose, its reason for being, all of which are seen to reside in the arrangement of its matter.

Treatises on Monsters: Fortunio Liceti

Almost any treatise on monsters that one examines dating from the early modern period will offer: (1) a definition of monsters, drawing from etymological analyses of the divinatory term *monstra* (and never, as we do, from a physiological definition of a "normal" human arrangement of parts); (2) a theory for the causes of monstrous births; ranging from excess or lacking semen and menses, contusions to the womb, constricted space, to the mother's imagination, and coupling with incubi; (3) a discussion on monstrous races, whether they exist or not, and what the author believes to be true; (4) a collection of anecdotes coming from the textual traditions (Pliny, Livy, Valerius Maxiumus, Aulus Gellius, Augustine, etc.), from more recent accounts (Lycosthenes, Paré, Aldrovandi, Rueff, Weinrich, Liceti, etc.), and finally from personal experience. Various classificatory schemes are used: uniform vs. multiform; excess vs. missing vs. misplaced parts; human-looking animals vs. animal-looking humans; and so on.

Of course, various constellations of beliefs are possible within a combinatory matrix. This is evident in a text which can be regarded as emblematic of seventeenth-century scientific monster treatises: Fortunio Liceti's *De Monstrorum caussis, natura, et differentiis.*[42] Liceti refutes the idea that monsters are to be interpreted as parts of speech following the rules of some prodigious syntax—God's way of speaking to his sinful earthly community—yet Liceti also believes very sincerely in God's ability to transform people into monsters in order to punish them, just as he takes for granted the real operations of malicious demons in transporting semen, exchanging embryos, and otherwise meddling in human birth processes. Liceti was a highly respected physician, intellectual, and Aristotelian philosopher. The same sort of seemingly contradictory combinations of beliefs can be shown in Ulisse Aldrovandi's *Monstrorum Historia,* Gaspar Schott's *Physica curiosa,* Thomas Browne's *Religio medici,* and Ambroise Paré's *Des monstres et prodiges,* just to give a pan-European sampling.

Belief patterns can follow complex trajectories and seemingly contradictory paths. To commit oneself to sorting out what is "scientific" from "archaic," "superstitious," or "religious" is not only a monotonous and unfruitful enterprise, it also distorts and renders opaque a more nuanced and interesting set of issues that are indigenous, so to speak, to the period under study. When I propose, then, that early modern scientific literature on monsters subscribes to a descriptive mode, I do not mean to imply that this is the only mode present in these texts. Nor do I mean to imply that adoption of a descriptive mode precludes other modes. The fact that description indicates a philosophical and ideological interest in the material conditions of being does not make the co-presence of other interests contradictory. What it does imply, though, is that within the combinatory matrix of a "period culture" (to use a convenient neologism) all the modes of belief take the same building-block concepts as their starting points: in this case, the basic concepts provided by Western philosophy of spirit, form, and matter.

As we have seen, when monsters are viewed as material products of defective nature they lose their terror factor and enter instead into a collocation of knowledge: from being abandoned on deserted islands, to being examined in fields, to appearing as decorations on hospital walls and ordered in glass jars on museum shelves. Their interest was not purely scientific, of course; the *wunderkammern*, or cabinet of curiosity, mixed up artificial and natural beings in provocative ways, precisely to titillate curiosity and wonder. But it is safe to say that museum monsters provoked interest for their philosophical origins as much as for their horrific appearance.[43]

Similarly, monstrous races, when viewed as products of excessive heat in faraway lands like Ethiopia and India,[44] and not as a decorative function in a *concordia discors*, were recuperated into a coherent scientific paradigm that explained abnormal human arrangements as being due to the struggle between hot and cold, male and female, form and matter. The resulting combinations were not terrifying, they were not even especially wondrous; they were simply one more "proof" of the theory of generation being put forward. The point I am trying to make is that even within a materialist paradigm, nothing is implied regarding the existence or nonexistence of a supernatural realm. In fact, Liceti remains faithful to the Aristotelian conception: a monster is the failed product of an intentional act to reproduce oneself, a failure displayed in the arrangement of members: "One calls a Monster everything that is born between animals with a disposition and an arrangement of members all different and completely contrary to the Nature of those that engendered them, such as, for example, a child without feet, a little girl with two heads, a child with a dog's head, a Centaur, and other similar ones."[45] Liceti thus clears teratology of all moral monstrosities and prodigious signification. The professor of medicine also dismisses minor deformations and mutilations, and monstrous races in Africa such as pygmies.[46]

Liceti proposes a threefold division of generation: *supernatural, infranatural,* and *natural* productions. And he adds that, even though he will only speak about the natural causes, he will do so without forgetting that "the sole, efficient cause is Almighty God, that is, motive Intelligence and the Heavens."[47] Now what is interesting about these broad divisions is how they filter down into the etiologies of particular kinds of monsters. While the miraculous workings of God must not provide fodder for human speculation, the workings of demons provide a whole chapter in Liceti's treatise. As for the workings of nature, even though Liceti speaks "properly and as a Physician," nature includes everything from the parents' imagination to out and out "grafting" of limbs between children in order to fabricate freaks for profit.

Just out of curiosity, let us look at Liceti's treatment of monsters with double natures, like the one Varchi described.[48] First cause: excess matter in the formation of one member, too little in the other. Second cause: narrowness of womb and defective formative power (the semen). Third cause: illness of the fetus, superfetation (double impregnation). Fourth cause: menstrual matter flowing into the uterus, either through inflammation due to the mother's imagination or simply from too much semen. Fifth cause: heredity, i.e., double-natured parents. The sixth cause lies in "the vehement imagination of the parents, and in a defect of nourishment on the part of the fetus."

The seventh cause—"artifice imitating the error of nature"—is especially intriguing, and in fact Liceti warns the reader that it might seem outlandish, especially since "in this work we primarily deal with physiology": "But since it is fitting for the philosopher to contemplate all the works of nature, whether she operates on her own, or whether she cooperates with another means in making some work, even by artifice, so this may happen in monsters since artifice cannot act without the help of nature, the work of art consisting solely in the application of active principles to passive principles of nature herself."

Let us file away this definition of the works of art as the application of active principles to passive principles for a later discussion. For now what I would like to focus on is the fact that this sort of intervention into nature's productions through human skill and inventiveness—likened, in fact, to the art of grafting plants practiced by the "most perspicacious master Tagliacotius"—when applied to the production of monstrous human beings, is branded as entirely abominable and barely acceptable in a respectable treatise of this sort. Liceti's apology that a philosopher is bound to contemplate all of nature's works stresses to what extent this sort of practice was viewed as contemptible.

And if we had any lingering doubts as to the desirability of artificially fabricating these sort of freaks, Liceti makes it clear what kind of people stoop to such low and venal pursuits: "Sicophantas," lying impostors, greedy parasites who take advantage of the credulity of the public:

> It is not unusual to see these wanderers touring the world to exhibit marvelous monsters for profit. To fabricate [conficio] them they first cut open the carnal parts of young children's bodies, like the back, the nose, the arms and stuck one part to the other. With nature's help and by transfusion of blood and nourishment the parts were able to

fuse into one. There remained only to amputate some other part to lend them a most horrible, monstrous appearance. May God preserve us from such scoundrels and may they be severely punished by our Princes.

Monster Recipes: The Manual of Natural Magic

Now in Liceti's (and presumably many of his readers') view, the production of living monsters through the application of artificial means to complete the workings of nature was in some fundamental way contrary to nature. The main reason such a practice was perceived as contrary to the normal functioning of the world is because at least since Aristotle's time art had been defined as imitating nature, and not the other way around: "Generally art partly completes what nature cannot bring to a finish, and partly imitates her."[49]

The Baroque taste for inverting this equation has been well documented; however, the mixing and blurring of the distinction between *naturalia* and *artificialia* would only have provided such delightful transgressive pleasure if crossing over these boundaries was in fact maintained as a transgressive act during this period. That is to say, nature was unquestionably believed to be the primary term, and human art, relegated to completing or imitating nature's works, was equally unquestionably believed to be the secondary term. The play, the fun, lay in simulating an illusion of reversal of these two terms.

However, if we leave the domain of teratological treatises and pick up another sort of text in which monsters frequently make an appearance, we are faced with a puzzling contradiction. If the production of living monsters was such an abomination to decent scientific thinkers, what are we to make of long and detailed instructions on this art in manuals of natural magic? Here we can turn to Giambattista della Porta's extraordinarily successful book *Magia Naturalis,* arguably the most popular and well-read manual of its kind in seventeenth-century Italy (the expanded Italian translation came out in 1611).[50]

It is important to note that although Porta's manual was written by a philosopher who was also a man of letters and a proponent of the *ars reminiscendi,* the art of memory, the "soiled and damaged examples [of his *Natural Magick*] testify to the fact that it was used and worn out by the practical man and in the laboratory rather than being safely preserved on a library shelf." Derek Price, in his introduction to the English translation I will be using, also insists that Porta's manual represents the collective work of the "first scientific society of our age." As Porta himself expresses the philosophy of the Accademia Secretorum Naturae of Naples, many of the experiments necessary "to know whether what I heard or read, was true or false, that I might leave nothing unassayed" were conducted by the group of "Otiosi" associated with him. The indefatigable researcher was, by his own testimony, never lacking "the Labours, Diligence, and Wealth, of most famous Nobles, Potentates, Great and Learned Men, wanting to assist [him]" (preface). Price confirms that the fact that "the work was done not by one man alone but by the Otiosi . . . makes it of more than passing interest; one can never dismiss it as a chance 'precursor' or as the product of a capriciously remarkable man. It is typical of its times and illuminates them trustworthily."

Porta provides twenty fascinating books full of recipes, spells, household tips, and scientific experiments that span all the arts of the modern magician. The Neapolitan takes care to distinguish the common conjurer, witch, or sorcerer from his magician, who is expected to be not only "a very perfect Philosopher" and "a skilful Physician": "Moreover it is required of him, that he be an Herbalist"; "and as there is no greater inconvenience to any Artificer, than not to know his tools that he must work with" he must "be as well seen also in the nature of Metals, Minerals Gems and Stones." "Furthermore, what cunning he must have in the art of Distillation . . . no man will doubt of it: for it yeelds daily very strange inventions, and most witty devices." "He must also know the Mathematical Sciences, and especially Astrologie," and finally, "the Opticks."[51] Natural magic, it would seem from this curriculum, bore little resemblance to what we usually associate with the diabolical crafts of witches and diviners, although the distinction was not crystal clear to at least one of his contemporaries who denounced Porta to Pope Paul V for "dabbling in the occult."[52]

The well-thought-out philosophical and historical overview on natural magic that Porta offers in book 1 draws heavily on Marsilio Ficino's *De Vita Triplici*, Agrippa's *De occulta philosophia*, on Neo-Platonic writers such as Iamblichus and Proclus, as well as on the ubiquitous writings of Hermes Trismegistus. While what I have called the *descriptive mode* rises out of a philosophical belief that spirit is a product of the arrangement of matter, thus leading to the view that monsters are most appropriately described, classified, and placed in some taxonomic system of knowledge, we will see that there is another contemporaneous tradition in which *spirit is seen as co-penetrating matter*, according to the Neo-Platonic tradition of the *anima mundi* or world soul. In this mode, monsters—as natural products subject to the skillful manipulations of the artificer—are most appropriately induced from matter and used for the benefit of humankind. This is what I will call the *manipulative mode*.

In order to understand seventeenth-century Neo-Platonism it behooves us to start our exploration of natural magic with Marsilio Ficino's wonderful third book on the triple life *De vita coelitus comparanda* (1489), a practical manual on how to bring one's life into salutary harmony with the celestial influences, or as he describes it, "our laboratory here . . . our antidotes, drugs, poultices, ointments and remedies . . . enclosed in the following book . . . offered as a kind of medicine for the powers of life."[53]

"Nature," writes Ficino, "is everywhere a magician," a shorthand expression to mean that the powers of nature, the attractions or passions, draw things to each other and repel them just as the magician "with a certain art, gathering many things into one, correctly and appropriately . . . can somehow draw down . . . heavenly things at the right times to men, making the lower things in agreement with the higher." In the same way nature "clearly entices certain things with certain foods, just as gravity draws heavy things to the center of the earth." These resonances or attractions between sublunary and celestial beings are due to the fact that the world is infused with a soul, emanating from the Divine Mind, and this live being, this huge creature that is the world, is actually "everywhere copulating with itself out of this mutual love of its own limbs. They say that it exists in such a way that the bonds that hold these limbs together are inside its own mind, which going through its limbs, works the whole mass and mixes with the great body itself."[54]

Such a magnificent, erotic animal, paradoxically generating itself out of its own mind—hopelessly confusing the ontological orders of matter, body, and mind—seems to share absolutely nothing with Aristotle's ordered world of potential matter, actualized in form. And yet, the categories of matter and form do survive intact in Ficino's vision: the *anima mundi* "divinely contains at least as many seminal reasons for things as there are ideas in the divine mind, and with these reasons it fabricates as many species in matter." Matter and form, as in the Aristotelian system, are inseparable, but with an enormous difference. Neo-Platonists view all matter-form combinations as being animated with spirit: "No one, therefore, should think that in certain materials of the world there are numinous elements, separated, inside of material, and that these elements get drawn out; but one should rather think of these as demons and gifts of the animate world and the living stars."

Even more alien to an Aristotelian is the idea that these "demons and gifts of the animate world" can be coaxed, lured, cajoled into changing their abodes by manipulating the arrangement of the matter: "No one, furthermore," explains Ficino, "should marvel that the soul can be, as it were, allured through material forms. If it can be allured to harmonious foods by these forms, it does so, and it always and freely dwells among them."[55]

For example, we can construct a gold talisman in the shape of a solar being, such as a rooster or hawk, and if we were to engrave an image of the sun on the bird simulacrum, with characters to signify Jove, and set a solar gem like a carbuncle or ruby into it, we could induce healthful solar influxes into the formed metal. In order to benefit from the solar spirit thus called into the talisman, we need only wear it around our neck, thereby allowing the celestial influence to penetrate the body.

Interestingly enough, Ficino compares these sorts of magic arts to agricultural arts: "One prepares a field and seeds for heavenly gifts, and with certain graftings one propagates the life of a plant, leading to another and a better species. . . . A philosopher learned in natural and astral matters, whom we call therefore a Magus, does the same thing, with certain earthly enticements drawing the heavenly things when he does it properly, sowing no differently than a farmer who is knowledgeable in grafting, who starts a new shoot off old stock."[56] This image of grafting cannot help but remind us of the art of grafting human limbs that Liceti described among unscrupulous, nomadic profiteers.

In both cases, what is being discussed is a particular relation between art and nature, and the role that human beings can assume according to the degree of their knowledge and skillfulness. Human matter that is formed in an aberrant shape, the Aristotelian would say, is simply due to the more or less random operations of nature in overcoming the resistance or deficiency of the material or in directing the movements of the semen. If the matter is perfectly formed in line with the intentions of its generator, then it is a perfect human being. By definition, the more it deviates in resemblance from its male parent, the more monstrous it is considered. Its status in the hierarchy of beings depends entirely on the arrangement of its limbs.

The Neo-Platonic view is different. "There is nothing so deformed in the whole living world that it has no soul, no gift of soul contained in it," says Ficino. "Does not nature, the creator of the fetus itself, when it affects the little body with a certain kind of arrangement it has, and shapes it, does it not bring forth spirit from the universe immediately with this preparation, right in the fetus itself, as if it were a kind of food it was giving it?"[57] The point is not so much that arranged matter is endowed immediately with spirit, directly from the universal soul, the instant that it is formed. What is really striking is that this natural, magical process is in every way available to human duplication. "Nature is everywhere a magician" intoned Ficino, but equally we might say that the magician possesses in all respects the powers of nature.

Art still completes nature's works, but the knowledgeable man can go much further: "Thus, the wise man, when he knows a certain material, or those partly worked on by nature, partly finished by art, gathers them even if they are scattered, and knows which heavenly influx they are able to receive. He gathers these with the planet reigning whose influx they contain; he prepares them, uses them, and obtains for himself, through them, the heavenly gifts."[58] He assembles particular materials, forms them through his art, takes the scattered pieces, and gathers them together in harmony with the cosmic forces. He *prepares, uses, obtains for himself,* what the heavens have to offer.

Manipulation is not a bad word. I could have just as accurately proposed the word *operation,* the word most favored by writers on magic and the magicians themselves. What appeals to me about *manipulation* is its connection with *manus,* the human hand. There is something distinctly anthropocentric about the philosophy of natural magic. Nature can "operate" just as man and machines and impersonal forces can, but only human beings have the capacity to manipulate matter and form. This is what I find so distinctive about these particular writings—and these particular monsters: they are created exclusively for the uses and pleasures of mankind.

Giambattista della Porta's Natural Magick

After this short detour into Ficino's Italian Renaissance version of Neo-Platonic magic, we are better able to understand the philosophical basis of Porta's sixteenth-century formulation that was the standard textbook well into the next century. The differences between the two thinkers do not lie so much in their ideas, however, but in the locus Ficino and Porta have their magicians occupy, and in the apparatuses that the operators employ.

Ficino's magus is counseled to "walk about a lot in the open air" as it "freely licks you . . . and as you dwell in lofty, pleasant places, penetrating you purely and marvelously presenting the world movement and vigor to your spirit." Costumed in appropriate white robes, the adept is advised to perform special dancelike movements in imitation of the celestial spheres: "Do this work so that you are turning in perpetual motion with these powers," accompanied by appropriate chants. "It is a good idea to choose the place that is the most fragrant. . . . I advise you to keep changing the location, although always with delight. . . . One should think of Paradise, and of the Tree of Life in Moses."[59] In a word, Ficino's magus belongs in the garden.

Porta's artificer, on the other hand, the ultimate experimentalist, is quite clearly ensconced with his instruments and tools in the laboratory. The frontispiece of the Latin edition (and others as well; see figure 2)

FIGURE 2

Frontispiece to Della Porta's *Magiae Naturalis* showing the magus at work in his laboratory. (Courtesy of Department of Special Collections, Stanford University Libraries)

actually depicts the author at work in his laboratory. Porta stands cloaked with hat in hand. In his left hand he holds a sword pointed at a mirror, which he gazes into as he places the tip on its surface. Between Porta and the mirror, the sun beams its effusions into the room, some of which pass through a chalice, poised on the edge of a table (or perhaps floating in the air? It is difficult to tell from the slightly distorted perspective of the flask). A stream of light passes under the chalice igniting a crystal or glass ball on the floor. In the foreground, before the fiery ball and Porta's outstretched foot, lie a quadrant and another mathematical instrument scattered on the floor. The complementary relationship between the natural force of the sun, the secret actions that Porta is performing with his art, and the placement and fabrication of the instruments (chalice, sword, sphere, and quadrant) is highly significant. The frontispiece portrays them all as simultaneously present and active in producing the desired effect.

Porta, having been the subject of the Inquisition's censure, is most adamant in stating that "the works of Magick are nothing else but the works of Nature, whose dutiful hand-maid Magick is." But immediately following this assertion is another one that more explicitly reveals human art as a technique to perfect deficient natural processes. The image used, once again, is an agricultural one: "For if she [Magick] find any want in the affinity of Nature, that it is not strong enough, she doth supply such defects at convenient seasons, by the help of vapours, and by observing due measures and proportions; as in Husbandry, it is Nature that brings forth corn and herbs, but it is Art that prepares and makes way for them."[60] The most important distinction to impress on his readers was not that between art and nature, a loose, time-worn one that allowed much latitude, after all; the main distinction to be made was between demonic arts—necromancy—and natural magic, "which all excellent wise men do admit and embrace, and worship with great applause."[61]

Porta offers a genealogy of these wise men, "the most noble Philosophers that ever were, Pythagoras, Empedocles, Democrites, and Plato," and then characterizes the "dark and hidden points of learning" that they cultivated as "the perfection of natural Sciences." By stressing that this body of knowledge is the most excellent and wonderful among all natural sciences, Porta places his work under the rubric of a praxis: "Others have named it the practical part of natural Philosophy, which produceth her effects by the mutual and fit application of one natural thing unto another." He also strives to strip this magic of any miraculous connotations. It may be a hidden art, full of occult precepts and properties, but there is nothing supernatural about it: "This Art, I say, is full of much vertue, of many secret mysteries; it openeth unto us the properties and qualities of hidden things, and the knowledge of the whole course of Nature; and it teacheth us by the agreement and the disagreement of things, either so to sunder them, or else to lay them so together by the mutual and fit applying of one thing to another, as thereby we do strange works, such as the vulgar sort call miracles, and such as men can neither well conceive, nor sufficiently admire."[62]

This passage provides two important ideas: (1) that the "strange works" of magical manipulations are only erroneously believed to be unnatural (and Porta turns his persecutors' fingers to point back at themselves: "Superstitious, profane, and wicked men have nothing to do with this Science; her gate is shut against them"); and (2) that this practice operates by taking things apart or putting them together in particular ways.

Liceti and those subscribing to the descriptive mode—Aristotelian types who enjoy defining, describing, and creating taxonomies of monsters that they find in nature—might be shocked by the contents of Porta's second book of natural magic, but those who think in the manipulative mode find it a perfectly acceptable, and indeed, desirable, pursuit. I am referring to the thirty-page section entitled "Showing how living Creatures of divers kinds, may be mingled and coupled together, that from them, new, and yet profitable kinds of living Creatures may be generated."

Typically for such manuals, the reader finds no definition of what monsters are, no reference to monsters as signs of any sort and little interest in monstrous races except as a conceptual goal for spurring on more experimentation. Anecdotes are offered purely to lend authority to the plausibility of producing such creations. And, as we would expect, theories on the generation of monsters appear not as a philosophical problem, as in Varchi, but in order to better understand exactly how nature operates, so that the magus can duplicate or imitate her processes more precisely.

"There are two generations of Animals and Plants, one of themselves, the other by copulation: I shall first speak of such as are bred without copulation; and next, of such as proceed from copulation one with another, that we may produce new living Creatures, such as the former ages never saw," Porta begins. He proceeds to discuss putrefaction, the processes of spontaneous generation of not only insects but other creatures as well, in the tradition of Empedocles' and Democritus's creation theories.[63] Frogs, mice, and lice are all able to be generated out of putrefaction; serpents, instead, may be generated of human bone marrow, the hairs of a menstruating woman, and from a horsetail or mane; and so on.

As promised, Porta next moves on to the art of carnal copulation: "Now we will shew, that sundry kinds of beasts coupling together, may bring forth new kinds of Creatures, and these also may bring forth others; so that infinite monsters may be daily gendred." In defense of those who are skeptical of hybrid breeding, and especially as to the reproductive capacities of second and third generations, Porta assures them that "we must not think that the one example of Mules not gendring, should prejudice the common course of other creatures. The commissions, or copulations, have divers uses in Physick, and in Domestical affairs, and in hunting: for hereby many properties are conveyed into many Creatures."[64] Clearly the production of "infinite monsters" has a primarily utilitarian function for Porta: physics, domestic chores, and hunting. Let us note also the complete absence of fear in regard to these artificial monsters.

It is one thing to fool around with producing various types of dogs cross-bred with tigers, lions, wolves, and such, since "pretty little dogs to play with" provoke little serious moral indignation, and various combinations of mules, bulls, horses, sheep, goats, and rams are in fact quite appealing to the imagination. (Who would not be delighted to have one of Porta's more exotic species in their *serraglio:* a hycopanther, for example, or a bactrian camel, "gendred of a Camel and a Swine"?) However, more dubious practices are to be found in chapter 12, "Of sundry copulations, whereby a man gender with sundry kinds of Beasts," and in chapter 17, "How we may produce new and strange Monsters"—practices such as interspecies copulation and mixing different semen in one womb.

These techniques naturally lead into a presentation of the various theories of monstrous generation, using Aristotle quoting Democritus and Empedocles, presented so as to buttress Porta's views. But the discussion on the ways nature goes about "fashioning of a living Creature" is only a prelude to the basic principle of this art: "And so, look how your art disposes and layes things together, and after the same manner, Nature must needs accomplish her work, and finish your beginnings."[65] The course has come full circle: the way the magician "disposes and layes things together" is the same way nature operates and not the other way around; nature imitates and finishes the artificer's works. The magician is really a kind of midwife to nature, setting up the proper conditions for her to accomplish her works.

Porta does offer some personal anecdotes: "I my self saw at Naples, a boy alive, out of whose breast came forth another boy, having all his parts, but that his head only stuck behind in the other boyes breast," he narrates. Unlike Liceti (or Paré or Aldrovandi), though, one senses his reticence and disinterest in curiosity-mongering. "Rhases reports, that he saw a Dog having three heads" he begins in the brief section entitled "monsters be generated in Beasts," but he immediately confesses that "there be many other like matters which I have no pleasure to speak of."

The reason he derives no pleasure from such monsters is because he had no role in *creating* them. Porta's genuine pleasure is restricted solely to the production of monsters: "For if you attempt likely matters, Nature will assist you, and make good your endeavours, and the work will much delight you: for you shall see such things effected, as you would not think of; whereby also you may find the means to procure more admirable effects."[66] This science of monster making, of producing "infinite monsters," is in fact predicated on its limitlessness, on its open-endedness. The delight stems from the manipulator's surprise in beholding the creative, unpredictable results of his tinkering.

Even more satisfying is the idea that the product, or the end, can serve as the means for the next experiment: one generates a monstrous hybrid, then breeds that monster with another, and so forth, with mind-boggling implications worthy of a Doctor Moreau. This is what delights Porta and his fellow laboratory men: that they are taking part in an endless manipulation of means and ends—ends turning into means and means into ends—in a vast, creative process of combining complex life forms in a partially unpredictable universe. Never before or since have men felt more godlike than when puttering around the early modern laboratory in their daily generation of infinite monsters.

Before leaving Porta and his fellow *Otiosi* to their experimentation, there is one other way of generating monsters that is pertinent to our investigation. It explicitly describes how one may form the living matter of human beings into special shapes after they have issued from the womb, in the same way fruits or cucumbers can be shaped by growing them in specially molded cases. Porta cites Hippocrates in reference to a race dwelling by the River Phasis that had very long heads. The Neapolitan believes that "surely Custom was the first cause that they had such heads; but afterward Nature framed her self to that Custome." His description brings to mind the practice of binding girls' feet in China: "The beginning of that Custome was thus. As soon as the child was new born, whiles his head was yet soft and tender, they would presently crush it in their hands, and so cause it to grow out in length; yea they would bind it up with swathing bands, that it might not grow round, but all in length." Eventually, Porta explains, "by this custom it came to passe, that their heads afterward grew such by nature."

A fascinating concept, that persistent human efforts—a sort of purposeful, artificially willed resistance of matter—can actually permanently alter natural processes over the span of several generations. This is where the science of monster making takes a very nasty turn, when this sort of genetic manipulation applied over several generations is applied to the butchering of live animals, who are then made to copulate, until the desired deformity becomes reproduced through nature: "So if we would produce a two-legged Dog, such as some are carried about to be seen; we must take very young whelps, and cut off their feet, but heal them up very carefully: and when they be grown to strength, join them in copulation with other dogs that have but two legs left; and if their whelps be not two-legged, cut off their legs still by succession, and at the last, nature will be overcome to yield their two-legged dogs by generation."[67]

Stories of Demons

One has the impression that monsters were highly visible in early modern Italy. Porta's recipe for two-legged dogs, repulsive by its very simplicity, arose in response to his having seen the mutilated animals on tour, "carried about to be seen," he says. We automatically think of other famous monsters who toured Europe, both in the countryside and at courts, documented in pamphlets, portraits, and personal accounts. "Lazaras, an Italian," the one Porta mentioned having seen in Naples, for example, was granted a six-month license in London in 1637 "to shew his brother Baptista, that grows out of his navell, and carryes him at his syde."[68] Or the group of four indigenous people who were brought from Greenland during a 1654 royal expedition, to be exhibited along with their hunting and fishing gear, and who died shortly after being converted to Christianity. No matter, since their portraits could still be hung and displayed in Christian V of Denmark's museum (see figure 3).[69] Ethnographic curiosities often extended to live specimens, like the six Indians brought back from America to Francesco I's court at Florence, along with an assortment of monkeys, parrots, and other New World exotica. They, too, died after a short time. Ferdinand of Tyrol's collection also included the portrait of a well-known traveling Italian giant named Giovanni Bona.

If you were unable to catch a live specimen on tour in the local piazza, or were not privileged enough to contemplate the dead ones, either in portraitures such as Bronzino's, or staring out suggestively from behind their glass prisons decorating the museum shelves and walls, you could always resort to acquiring a stuffed version, such as the "fake basilisks . . . sold in public squares by charlatans" mentioned by Count Lodovico Moscardo in his *Museo* (1656).[70] Adalgisa Lugli reports that these artificial hybrids—made out of various real animal parts pieced and sewn together to represent fantastic creatures such as hydras, basilisks, and dragons—were in great demand, much like false relics that were trafficked during the Middle Ages.[71]

Taxonomy and taxidermy seem made for each other in this period: both fabricate specimens, and both put the specimens on show for profit. Likewise, the natural magician was concerned with producing monsters for utilitarian purposes. His laboratory invited collaborators and visitors, en-

FIGURE 3
Greenlanders brought to Christian V of Denmark's museum as live
ethnographic specimens in 1654, portrayed with their hunting instruments
in B. Valentini's *Museum museorum*, 1704–14.
(Reproduced from Adalgisa Lugli's *Naturalia et Mirabilia*
[Paris: Éditions Adam Biro, 1998], p. 81)

couraged experimentation and verification, sought an exchange of information and results. These visible monsters strike us by their very corporality. Corporality intrudes awkwardly into the Varchi prose description of the Rucellai dissection, as we have noted. When reading the Porta recipe for creating two-legged dogs one wants to avert one's eyes, dissociate from the reality of what those crude words instruct: "We must take very young whelps, and cut off their feet," "join them in copulation with other dogs," "and if their whelps be not two-legged, cut off their legs still by succession."

As fully packed with visible monsters as early modern Italy was, it was equally dense with more elusive, less corporeal monstrous beings that, although mentioned extensively in the literature, one is never really able to locate in the ways we have been able to locate monsters in the garden, the museum, the piazza, and the laboratory. The reason is because the third kind of literature we will examine deals with forbidden, illicit, occult kinds of monsters, neither corporeal nor incorporeal, but made of a finer substance than ours. I am referring, of course, to demonology texts.

Demons and monsters are linked in two distinct ways. One linkage is based on an obvious and relatively unproblematic correlation between moral depravity and corporeal deformation whose association is widespread and deeply ingrained: the demon assumes a monstrous appearance; or, to be more precise, demons *are* monsters, monsters *are* demons. A typical anecdote from Francesco Maria Guazzo's *Compendium Maleficarum* of 1606 illustrates this association in a marvelous fablelike anecdote:

> About the year 1520 at Basle a certain simple-minded tailor afflicted with a stammer somehow or other entered the cave which has its entrance at Augst. . . . [A]fter burning consecrated corn . . . he first went through an iron gate, and then from one room to another, and so into a most beautiful flower garden. In the midst was a magnificently decorated hall where he saw a very beautiful Virgin with a golden diadem upon her head and her hair flowing loose; but the lower part of her body ended in a horrible serpent. This Virgin led him by the hand to an iron chest which was guarded by two black Molossian hounds, which by their baying prevented any from approaching: but the Virgin in some wonderful manner calmed them, and taking a bunch of keys from her neck, opened the chest and displayed every sort of gold, silver and copper money. . . . He added that the Virgin said that she was a King's daughter who had always led a devout life, but that she had been changed by an evil spell into that monstrous shape and that there was no other hope of her recovery than for her to be kissed three times by a youth of unimpaired

chastity; for then she would be restored to her former shape, and would give as dowry to her liberator the whole treasure hidden in that place. He asserted also that he twice kissed the Virgin, and twice saw upon her face such a horrible gloating at the thought of her liberation that he was afraid that he would not escape from her with his life. Afterwards he was taken by some loose boys into a brothel, and could never thereafter find the approach to the cave, to say nothing of entering it again.[72]

The siren figure is a rich, complex image which will be treated in detail in a following chapter on metaphor. Her association with lust and money is hardly fortuitous: most appearances of demonic monsters involve some kind of temptation, some beckoning to a loss of control, to a giving way to the animal, blissful, unconscious, purely corporeal sides of our nature. Christ's temptation in the desert, and Saint Anthony's encounters with a centaur and a dwarf on his way to meet Saint Paul (recounted by Saint Jerome) are the Christian textual prototypes for this topos.

This story has an unreal, otherworldly quality about it that seems diametrically opposed to the "descriptive" or "manipulative" prose styles. Guazzo's prose is loaded with adverbial and adjectival phrases, making it a very different kind of description, one that we would associate with a dream transcription or a fairy tale. The "simple-minded tailor afflicted with a stammer" has none of the ontological solidity that we experience in the subjects of Varchi's or Porta's texts. Rather, his presence strikes us as a purely functional one in the narrative. He fulfills the role of protagonist, playing his part in the sequences set into place by the imperatives of the narrative. Consecrated corn, a cave, iron gates, a series of rooms, a garden, and then a magnificent hall; before the youthful hero and the beautiful princess stand two hellish hounds; keys, an iron chest, treasure; the helpless virgin suffering under a spell, the promise of a reward for setting her free—the whole account reads like one of Vladimir Propp's fairy tales.

Or does it? The reader's spell is broken abruptly when the hero kisses the virgin: "a horrible gloating" disfigures her beautiful features, instilling in him a fear for his life. Up until this point the fact that "she had been changed by an evil spell into that monstrous shape" seemed almost fitting: we unthinkingly classify the virgin under the rubric of "mermaid," bare-breasted hybrid of many an erotic tale. This moment of disquietude, of uncertainty prevails, the diabolical illusion is shattered, and the tale quickly ends in sordid, anonymous sex: "Afterwards he was taken by some loose boys into a brothel." Chaste no longer, the simple-minded tailor is henceforth barred entrance to his fantasy cave: he "could never thereafter find the approach to the cave, to say nothing of entering it again." A Freudian reading would point to the Medusa association ("her hair flowing loose"), the phallic lower body of the serpent-virgin, the fact that the tailor, in spite of his stammering, is somehow able to enter the cave, the gender roles, and so on. We may briefly note that these confessional fantasy monsters remain exceedingly mercurial, elusively establishing a purely textual or narrative presence as characters in fantastic tales. There are hundreds of such accounts to browse through.

The other way that demons and monsters appear together in demonology texts is that demons are said to be responsible for parenting monsters. While demons are unable to actually provide their own semen (this would violate the precept that only God can create life; also, being formed out of condensed air or vapors, demons are unable to produce their own semen), the diabolical beings do have at their disposal many techniques of "artificial insemination." Liceti provides a comprehensive list.[73] Illusion, incantation, substitution, contamination, deformation: these are some of the diabolical techniques available to enterprising demons. Apart from the first—hallucination under bewitchment—all the practices that Liceti describes fit into the standard Aristotelian schema of generative causes, derived from the interaction of formative virtue and passive material. Demons do not actually invent or create anything; they simply help along or impede natural processes, either by interfering with the female matter so that it is unable to be properly formed, by diluting the semen so that it is unable to overcome the resistance of matter, by blocking nourishment to some of the fetus's body parts, by provoking an illness, or by suggesting strong images to the parents' imagination. These processes all fit into the physician's normal etiology of monstrous births, which the demon simply enhances in one way or another in order to produce an abnormality.

A standard event in demon stories takes place the instant the subject who has conjured the diabolical forces begins to suspect that he or she has been duped. The virgin-serpent's gloating face which tips off the tailor to her "true" nature is emblematic of dealing with beings from the *infranatural* realm. One of the most perplexing and anxiety-provoking problems related to demons was how to tell the good ones from the bad ones, since demons and angels were similarly constituted, created by God at the same time, and both inhabited the sublunar spheres immediately enveloping the earth. This easily reversible duality between good and evil demons—in fact, the Greek word *daimon* originally meant "divine being"[74]—maintained in place through the devil's mastery of illusion and trickery was what made illicit magic so dangerous to the practitioner.

Conclusion

I will bring this survey of scientific monster literature to a close by pro-
posing one more relation between spirit, matter, and form that I believe
underlies these perilous, illicit arts. This third formulation is much harder
to characterize since the literature intentionally hints without telling.
Just as the creatures involved in sabats, hallucinations, and visions have
a flickering corporality and an uncertain ontological status—as we have
seen they tend to be purely textual or narrative creations—so is their
presence in texts somewhat elusive. One must look more closely, infer
more, make more tenuous connections in order to "bring them into view."
In this third mode—which I will characterize as coercive—*pneuma or spirit
is called into matter or forcibly inhabits it.*

Coercion of spirit into matter is in fact the monstrous mode par excellence since it is the only one that retains an association with the sacred,
right up to our own day. The descriptive mode, the belief that spirit arises
from the arrangement of matter, develops into scientific appropriation of
deadened, classified specimens that are firmly placed in a network of
"objects" and thus unable to assume any sort of "subjecthood." The deformed fetuses studied by comparative embryology, reproduced in textbook illustrations, provoke anything but terror. They speak clearly of
failed matter, botched creation, deficiencies in the birthing process.

The manipulative mode, the belief that matter and spirit co-penetrate,
is equally uninducive to conceptions of sacred monstrosity, because monsters permeate all natural processes in the form of creative hybridization
and random results. The more monstrous the creation, the more powerful
and delighted the magus-experimentalist feels.

In the next chapter I will argue that the terror and danger associated
with monsters are only present when there is some question of overstep-
ping the limits of creative manipulation. When manipulation becomes
coercion, when the practice of cutting up and breeding animals in order to
perfect nature's work is extended out of the laboratory, for instance, and
into the ideal republic in the form of planned eugenics to produce a perfect
race, that is when the monstrous rears its hideous head. Monsters and
technics go hand in hand.

THREE

Monstrous Machines

In the last chapter, I proposed that monster literature in early modern Italy can be fruitfully perceived according to three modes which correspond to underlying assumptions regarding relations between spirit, matter, and form. This heuristic device has proved helpful in answering the question: What happened to sacred monsters during the rise of the materialist paradigm? Did they really undergo accelerated secularization and lose their ancient association with divine forces, thus succumbing to the widespread extermination of supernatural beings that took place during the century which gave rise to the Scientific Revolution?

The answer has proved complex and has radically transformed the question. For one thing, secularized monsters—those that I characterize as specimens or objects of taxonomic description—have always existed, in physiological and zoological treatises as much as in theological considerations of monstrous races. Secondly, an association with the sacred cannot always be characterized as a religious one. Angels and demons are supernatural and infranatural beings, but they also serve as what we would call natural forces: gravity and magnetism, for instance, can be coherently described according to the precepts of natural magic as interactions between various kinds of spirits. Furthermore, although some monsters are figured as diabolical beings, not all demons are monstrous. Nor do "religious" thinkers necessarily perceive bodily deformation as a metaphysically meaningful sign. Materialist and spiritualist philosophies can combine in all sorts of ways without seeming to threaten the resultant system with incoherence.

Far more useful, then, is an analysis of what makes certain monsters sacred and others apparently not at any particular time. The categories furnished by Western philosophy of spirit, form, and matter provide this understanding. In my opinion, there is at least one continuity in what constitutes monstrosity throughout Western culture. From the earliest written records to present day, a necessary condition for defining a sacred monster is *that which is inanimate yet moves of its own accord*. Or, in the terms I have proposed: whenever spirit is called into or forcibly inhabits formed matter, there is a danger of monstrosity arising. This formulation of spirit, form, and matter, although not sufficient to describe a monster, constitutes a sacred view of the phenomenal world.

By this I mean that there are other kinds of sacred monsters as well, those that embody other kinds of transgressions of "natural laws," but this particular definition serves to trace a notion that is continuous from the pre-modern, to early modern, to modern eras. Furthermore, other sacred monsters, defined physiologically (abnormal bodies) or socially (scapegoats, as in René Girard's formulation), evolve into forms that are difficult, if not impossible, to identify under one category.[1]

Another source of complexity in assessing what happened to monsters during this period arises from the interaction of "real" or bodily monsters with hypothetical and textual ones. This distinction becomes even more problematic when we focus on man-made physical monsters, designed and crafted according to the architecture of the imagination. Placing odd creations of nature in a traveling freak show or an ethnographic collection fits well within what I call the descriptive mode: a secular interest in observing material deformation, based on the assumption that a being is monstrous because of the arrangement of its parts. This ludic event, an organized spectacle for the sake of curiosity, might be easily confused with the equally ludic and therefore apparently secular display of monstrous automatons or mechanical moving sculptures in private and public museums.

But what makes an automaton monstrous is not the arrangement of its parts (although the automaton is often formed to represent a monster, a highly significant convergence). That is to say, the disposition of its limbs is not what makes it rare and extraordinary; that is not what makes it a *monstrum*. Rather, it is the fact that matter formed by artificial means and moving of its own volition would seem to be endowed with spirit.

According to my cognitive schema, then, the sacred monster did not die out; it transmuted and migrated into mechanical contrivances.[2] The horror and fear provoked by appearances in nature of monstrous births moved over into the horror and fear provoked by our own artificial creations, where these affects have remained lodged to this day. The idea that there are laws in nature that humankind can perceive, formalize, and then utilize to go beyond nature's capabilities is also the basis for fears of transgressing some other law or imperative. Whether the other law that is perceived as being transgressed is an ethical or a mystical one is difficult to say. Nevertheless, an association with sacred retribution is strong: for example, the (well-founded) fears of ecological catastrophe that are projected onto apocalyptic visions of the earth taking revenge; Philip K. Dick's cybernetic nightmares, in which human and robotic consciousness disturbingly mingle and lose their distinctiveness; and the intersection of horror films with science fiction scenarios, such as *Aliens, X-Files,* and so on. In the seventeenth century this fear was expressed in terms of licit versus illicit magic: good magic operated by applying natural forces to one another, whereas bad magic obtained infranatural or diabolical aid. This distinction was never more crucial than in the art of making monsters, both organically produced hybrids and mechanical simulacra.

What I will offer in this chapter can only be a partial and highly directed tour through *seicento* museum collections and technological artifacts. My interest is clear: to show that monstrosity and technical virtuosity tend to appear in the same discursive contexts, and then to probe what the relation is between these two seemingly unconnected terms. Why do they appear together so often? And why do monsters appear so often in museum catalogs, rather than in the prodigy books of the previous century?[3]

One of the points I would like to make is that monstrosity is redefined in different times and places, so that if we insist on clinging to one definition we will be unable to perceive the monster unless it conforms to that definition. Certain characteristics of the monster are perhaps universal, but only ones that describe the place that the monster occupies in the social order and its relation to the interpretive community that defines it as such. In other words, a monster is always an indication of transgression, of breakdown in hierarchy; it is quintessentially a symbol of crisis and undifferentiation. However, beyond these attributes a monster can take on any form: a birth defect, twins, an albino, a witch, an extraordinarily talented person, an extraordinarily depraved person, a robot, the state, tyranny, communism, and so on.[4] What I ask of my readers is precisely the willingness to abandon whatever is their most evident representation of the monstrous (Frankenstein, Hitler, Extraterrestrials) and to allow the category to stray just a little, so that we may stretch it out and come to understand its operations more fully.

Perhaps the best way to approach this chapter is to understand it as an extended meditation on a comment Augustine made in regard to Egyptian practices of creating false gods by calling demons into statues: "as if there were any unhappier situation than that of a man under the domination of his own inventions."[5]

Planned Parenting

One of the most striking differences between Ficino's fifteenth-century magus and Porta's seventeenth-century natural magician is the inclusion of mechanical arts into the artificer's repertoire of skills. Ficino's adept fits himself to the seven celestial modes through seven grades of things: (1) images, (2) medicines, (3) vapors and odors, (4) musical songs and sounds, (5) gestures of the body, such as dancing and ritual movement, (6) concepts of the imagination, and (7) philosophical contemplation.

As we know, the artificer has little time for such contemplative and aesthetic pursuits. He is busy working in his laboratory to unravel the secret operations of nature, stuffing live geese into boiling water, sending up flying dragons, building furnaces for coloring plates, extracting aqua vitae, investigating the effects of magnets, fabricating damask knives, and preserving apples in sawdust. The titles "The increasing of Houshold Stuffe," "Of changing Metals," "Of the Wonders of the Loadstone," "Of Beautifying Women," "Of Distillation," "Of tempering Steel," "Of Cookery" are a sampling of the books that constitute *Natural Magick*. Porta's renowned book 17 on optics—"Of Burning-glasses, and the wonderful sights by them"—was immensely influential, quoted and emulated by virtually every successor.

We might say, then, that the experimental program sketched out by natural magic in this period had a solidly technical bias to it. Not surprisingly, the presence of technical instruments and procedures in texts subscribing to this mode also belied an implied interest in production and usefulness. As we noted, the "daily gendering of infinite monsters" is at least purported to be for the purposes of "physics" (that is, in the interest of experimental furthering of knowledge), "domestic chores, and hunting."

The reader is hardly unprepared, then, to find that in the same book ("On the Generation of Animals") that so copiously and thoroughly describes how to create artificial monsters we are also privy to Porta's instructions on how to improve human infants. Just as two-legged dogs can be educed from nature, human babies can also be manipulated and improved: "By some such practise as you heard before, namely by handling, and often framing the members of young children, Mid-wives are wont to amend imperfections in them; as the crookednesse or sharpnesse of their noses, or such like."[6]

But why stop at simply correcting undesirable features? By judicious use of the imagination and other experimental techniques practiced during conception and pregnancy, one can actually engender a male or female at will, and even produce perfect progeny according to whatever specifications the parents have in mind. Porta first explains how "to bring forth party-coloured Sheep" and horses, white Peacocks, and shag-hair'd dogs: "By this which hath been spoken, it is easie for any man to work the like effects in mankind, and to know how to procure fair and beautiful children."

> There was a woman [Porta recounts from personal experience] who had a great desire to be the mother of a fair Son. . . . [S]he procured a white boy carved of marble, well proportioned every way, and him she had always before her eyes: for such a Son it was that she much desired. And when she lay with her Husband, and likewise afterward, when she was with child, still she would look upon that image, and her eyes and heart were continually fixed upon it: whereby it came to passe, that when her breeding time was expired, she brought forth a Son very like in all points, to that marble image . . . as if he had been very marble indeed.[7]

This anecdote reads like an inverted version of the Pygmalion myth. The woman's desire is so steadfast and intense, so exclusively focused on the perfect statue that constitutes a simulacrum of her yearned-for object, that she achieves the replication and brings it to life. The son resembles the marble image so closely that he could have "been very marble indeed." The erotic fascination that Pygmalion's "ivory girl" exercised over him is displaced in this experimental account: the woman lies with her Husband, but "her eyes and heart were continually fixed" on the statue. Pygmalion's fetishistic desire is thus channeled into a more socially acceptable object.

Yet there is also a great difference between the stories: what is lacking in the early modern version is the intervention of any Venus as deus ex machina. The king of Cyprus makes an offering and prays that "my wife may be— / (He almost said, *My ivory girl,* but dared not)— / One like my ivory girl," and when his wish is granted he "is lavish in his prayer and praise to Venus, / No words are good enough."[8] On the contrary, this tale of procreation is only about human industry and success. A man will never be able to give birth, hence the tragic nature of the artist's desire and the need for miraculous intervention from the gods. Here, instead, the woman, empowered by scientific knowledge passed on to her by the male scientist (Porta says "after I had counselled many [women] to use it"), sets up the conditions for producing her desired object and, singlehandedly, using her own mind and body as instruments, attains her wish.

The happy ending of this tale of planned parenting makes the presence of chapter 20 ("How it may be wrought, that Women should bring forth fair and beautiful children") seem innocuous. "And thus the truth of this experiment was manifestly proved. Many other women have put the like course in practice, and their skill hath not failed them," Porta adds, thus emphasizing the active role of the mother's own skill and the complete absence of the father. But the woman literally makes of her body a laboratory, turns her procreative capacities into a scientific endeavor whose results make up the experimental data for confirming Porta's hypothesis. And perhaps we should remember that this is just one of many "experiments of living Creatures, both pleasant, and of some use."[9]

Whatever ambiguous benefits such experimental eugenics might have offered the woman in Porta's version—empowering her, allowing her to play the scientist, yet objectivizing and instrumentalizing her own body and imagination—are absent in Porta's pupil and successor, Tommaso Campanella. Campanella wrote his own manual of natural magic, *Del senso delle cose e della magia,* in direct response to Porta's urgings to provide a philosophical account of the workings of sympathy and antipathy. After the standard definition and noble history of natural magic, followed by an account of how its name has been defiled by being used in reference to demonic arts,[10] Campanella expresses his admiration for and debt to his two Neapolitan colleagues, Porta and Ferrante Imperato. He also specifies that his work provides what theirs lack: a philosophical accounting as well as a praxis.[11] "This knowledge, then, is speculative and practical both, because it applies that which it knows to works useful [*utili*] for the human species."[12]

Usefulness also provides the criterion for distinguishing the productions of natural magic from false miracles (*finzioni inutili*) proffered by charlatans: "And if you ask how I will distinguish true miracles from false ones, I answer that all miracles that are useless to the good of man are false, like that of conjurers, and the useful ones are natural artifices known to wise men."[13] All technological innovations, in fact, are classified as magic by the vulgar ("the invention of powders for firearms and printing presses was a magic thing, as was the use of the magnet"[14]) until their functioning is explained and their prodigious quality subsides. Clearly, natural magic, technical creations, and utility are inextricably connected in seventeenth-century texts. This relation is especially evident, in consequential ways, in regard to the generation of monsters.

Pregnancy was a very risky business in early modern Italy, but not only because of the high mortality rate. Although that may have been a concern, what really made procreation such a dangerous and tricky process was the facility by which the mother's imagination could influence the formation of the fetus. Now, this had some positive consequences, as we have seen in the inverted Pygmalion anecdote: by diligently harnessing her imaginative faculty, an enterprising mother could patiently mold her dream child. But the negative side was a much more common occurrence.

Everyone knows how capricious a pregnant woman is, and how subject she is to cravings. The reason for this belief, still with us today, lies in the extremely potent effect of these cravings on the unformed fetus: "Lo and behold the pregnant woman, yearning for ricotta cheese, touches the face and the ricotta is born on the baby." Desire—or the desiring spirit, to be more exact—acts on the unformed mass of matter in the woman's belly, and thus reproduces an exact image of itself, because "wherever an idea passes over, it makes its similar."[15] This is unfortunate, because even the most beautiful and chaste women can end up with unspeakable abortions or monsters solely from the effects of a wandering, undisciplined imagination.

Campanella does not mean to suggest that there are not other (Aristotelian-materialist) explanations for these monstrous births, especially if the astrological conditions are in sympathy, but imagination remains the most potent factor: "Beautiful females often produce monsters and Ethiopians [i.e., black-skinned peoples] solely from their imagination, although at other times, because of a malady of the semen, or a menstrual infection, or because of superabundance or lack of material or active virtues, the generation is tempered with combinations of monsters or beasts, especially if the stars of that combination tend toward like heat."[16]

In fact, Campanella rebukes the "foolishness of Aristotle" for assigning the active formative role to the motion of semen "since it would have to be very active and image-making to be able to express the image of the child."[17] This part of Fra Tommaso's theory goes back to Hippocrates: both male and female sperm combine with the effects carried by the semen-spirit (and later the imagination of the mother) to render the fetus. If the two semen are incompatible, "they make double men." The wild-card in this process is the celestial influences of the planetary and stellar effusions "because the stellar heat infuses with the particular heat of the beast,"[18] a side effect of the cosmic *consensus* between all things in the universe. All these factors, once systematized, go into Campanella's grand plan for the production of the perfect human race.

Several passages, both in this text and in his better known *Città del Sole* (City of the Sun) express the author's genuine astonishment and perplexity that the science of breeding and the intricacies of generation described above are not used for the "benefit of humankind": "It amazes me that we are so bestial that we neglect human generation but we place so much interest in the race of animals." The fact that we do not apply our knowledge to enhance our humanity is a sign of our primitiveness. We should get with it, keep up with technological innovations, apply our state-of-the-art knowledge of breeding to the progress of the whole community. His state-instituted eugenics program is an integral part of the ideal republic:

Therefore, in this Republic we should take measures . . . that marriages are accounted not according to dowries, but according to valor, or to couple worthy women with worthy men and make them gaze at statues and pictures of illustrious men of arms and letters and have them fall in love with them, and wait for the time when favorable stars are in the ascendant and in the middle of the sky, and at their most powerful position because they dispose and aid the nobility of progeny. And make sure that for everyone . . . to avoid generation through lack of menses, that the uterus is tight and does not soften the semen with menses, and afterwards at midnight, having digested, because the semen has grown, that they do not cause damage to the fetus by disturbing the spirit from its nourishment and diminishing it.[19]

This is only part of the plan. According to their character or temperament, certain people should never procreate at all: priests, for example, and philosophers who spend their time in contemplation. All their noble spirits conglomerate in their brains, so that their progeny are particularly bestial, corporeal, vulgar: "Be silent, Luther, who gave wives to priests, since he, too, had children who were rude beasts."[20]

The social aspects of Campanella's program are developed more fully in the *Città del Sole*. Here we learn that all sexual intercourse should be regulated by a medical doctor and an astrologer. Hygiene, control, and scheduling are the most important factors. The appropriate couples shall be selected by examining their genitals during public nude wrestling contests: "Being well bathed, they gave themselves to coitus every three evenings; and only big and beautiful females couple with great and virtuous men, and fat women with thin men, and thin women with fat men, to bring about temperings."[21]

Nothing could be further from Porta's experiments to produce creative hybrids and imaginative anomalies than the vision proposed by Campanella. Monsters are not only unwelcome signs of failed or misapplied techniques, they are positively dangerous for the stability of the community. Generation is rigorously and severely policed by the republic: repeated sodomy is punishable by death;[22] couples who engage in intercourse or acts of impurity within three days of conceiving are deemed to have sinned; all decisions regarding coitus must be ratified by the "highest master, who is a great physician, and stands below . . . the official prince."[23]

The result of all this care? The reason this eugenics program is so important? Because it produces an ideal race in the form of clonelike, civic-minded members of the republic, conceived under the same constellation, endowed with the same physical and moral characteristics: "di virtù consimili e di fattezze e di costumi." Cloning, the systematic exclusion of difference, especially of anomalies like monstrous births, is at the basis of civil harmony: "And this is stable harmony in the republic, and they love each other greatly and help each other."[24]

Campanella's breeding program is based on a well-reasoned philosophical rationale. Monsters are a product of nature's deficiency ("particular nature errs in making monsters through impotence or ignorance"), and deficiency "is the same as lack [*mancanza*]," which is the same as nonbeing, the condition of self-destructiveness which lies at the root of sin: "the quiddity of its sin is in us, who lean toward nonbeing and disorder."[25] We human beings tend toward disorder and self-annihilation ("the sinner is said to annihilate himself").[26] Our moral deficiency is the same lack that is mirrored in nature's deficiency. Thus, sin and monsters are the same thing; they both tell of chaos, destruction, nonbeing, annihilation. Campanella's strategy is simple: annihilate the monster before it annihilates you. Through order, hygiene, will, and control, the monster will be unable, literally, to come into being. Thank God for science and technology.

The technological basis of Campanella's utopia, like More's and Bacon's for that matter, is generally associated with a notion of progress (scientific progress especially), rationalism, natural religion based on natural law, an egalitarian society in which rich and poor are leveled, the sexes are granted equality, and everyone works together to improve the human condition. The oft-quoted comment by the Genovese sailor in the *City of the Sun* that "there has been more history in the past hundred years than the world has had in four thousand; and more books have been made in these hundred years than in five thousand; and stupendous inventions like the magnet and printing presses and firearms are great signs of the unification of the world"[27] is taken as an indication of Campanella's forward-looking commitment to the positive effects of technological innovation on social organization, especially when combined with his efforts to actually institute some populist version of the Solar Republic in his home province of Calabria.

I would opine a somewhat different view. The kind of technological thinking that Campanella espouses keeps intact the esoteric figure of the Artificer as God, not Porta's exoteric artificer, who shares his knowledge—precisely in order to convert more members to aid him in his experiments and to endlessly multiply and mutate the number of creations, turning ends into means and means into ends—but rather the Artificer as exclusionary Controller, as the Great Manipulator.

Listen to Campanella's version of the inverted Pygmalion myth: "And whoever knows how to reason about plants and animals will make children be born according to his own fashion, and he will uplift or debase human generation with these arts. . . . Now, whoever knows how to instill desire for a large or small thing in the pregnant female, can make the baby born with the shape and colors of that thing. . . . This affection brings the child out and it becomes exactly what was sought for by the magus."[28] The personal pronoun *whoever* (*chi*) in this statement does not refer to would-be mothers interested in learning how to beautify their offspring. It does, however, refer to any magician who is interested in employing a pregnant woman for his experimental ends. The real art lies in manipulating the desires of the pregnant female, who in this model becomes nothing more than an instrument or tool of the male magus.[29]

Artificial Magic, Artificial Monsters

My intention is not to vilify Campanella. What I am trying to point out is the connection between theories of monstrous generation, attitudes toward technological products, and the role of the technologist or artificer. Up until now I have organized monster literature under three categories, depending on which mode it subscribes to. Now I would like to propose two broadly based attitudes toward monsters that I think will be useful in pursuing the relations between monstrosity and technics.

The *descriptional* attitude adopts a passive stance toward monsters. Literature written in this stance generally creates a rubric in an encyclopedia, and places monsters somewhere in its taxonomy, or it characterizes them as *discors* in a global *concordia*. It also transcribes demon stories, describes birth defects, or interprets monstrous births as warnings from God. The other attitude, which I will call *transformational*, adopts an active, interventionist stance toward monsters, which it describes as being induced or created for licit or illicit purposes. These include organic creatures, produced through breeding, grafting, and other generational techniques, and inorganic monsters created by technological wizardry or through coercive means by calling spirits into inanimate matter.

The transformational attitude and the way it combines organic, inorganic, and demonic monstrous creations is clearly present in a passage from *Del senso delle cose e della magia* which closes book 4, chapter 18, "Magia della Generazione." The question under discussion is whether man can not only influence generation but actually create, or bring into being, a life form. Campanella begins by allowing that the creation of plants and imperfect animals generated through putrefaction, such as fleas and serpents, is well within the capabilities of a magician's art, but certainly not the creation of perfect animals, like horses, elephants, and men. He then examines all possible contenders for artificially created beings and refutes the possibility that they were brought into existence purely through the arts of natural magic.

First, the famous automatons of antiquity, Archytas's mechanical dove, Daedalus's moving statues, and Albert the Great's infamous talking head. Secondly, he discusses the existence of satyrs and other strange hybrids reported in faraway places. The next case to be examined is speaking statues, which must be inhabited by demonic spirits (and not human or animal ones), he reasons, because these are the only kind of spirits capable of being trapped inside metals. Finally, he proposes that satyrs are nothing but "uomini selvaggi," uncivilized men grown out of wild boys who were lost in the woods during childhood.[30]

The fact that monstrous races, New World savages, mechanical birds and statues, speaking idols, and spontaneously generated animals are all grouped together in the same context is not fortuitous. The fact that Campanella treats all these phenomena together as correlated practices or artifacts indicates that in some aspect they were indeed seen as equivalents. The one characteristic that unites savages, automatons, and speaking statues is the fact that their parenthood is in question. Savages and monstrous races seem not to be descended from Adam; hence there is a question of whether they were born directly from Mother Earth, as Lucretius described. Animated statues would appear to be self-generated, which is why Campanella brings up their exquisite appearance: the idea that "such beautiful statues" would spontaneously appear in nature out of nothing, with no generator but themselves "seems too much to believe." As for the famous automatons of antiquity, although their likeness to creatures of "spirit and flesh" and their ability to fly alongside real birds would convince some of their prodigious nature, Campanella is quick to demystify their functioning: "This is done with counterweights, like clocks, which, winding up some wheels spun with wire, raise and lower their wings."

Technological creations, monstrous races, and demonic idols are often grouped together in early modern Italian texts, not just here in Campanella's musings on the possibilities of generation. I have already mentioned that Porta's manual, a well-used laboratory guide, is full of technological processes and products alongside his magic of generating monsters. Campanella's manual picks up on some of the same experiments and adds his own. Many of these artifacts are for transforming oneself or others into monstrous beings. For example, "the whole face may seem various and deformed": "When the Glass is once made plain, put it into the furnace again, and let it be turned by the skilful hand of an Artist, till it lose its right position, then foil it."[31] Through the use of concave lenses, one can simulate the obscene deformations of a Bacchus, or a Priapus.[32] Or one "may seem to be divided in the middle" or "seem like an Ass, Dog, or Sow." What all these recipes and apparatuses have in common is that they are used to *transform* the human into the monstrous. For example: "By boiling an ass's head in oil you will make all the bystanders appear to have an ass's head."[33]

The transformation of bodies is intimately connected with the transformation of human beings into gods, and specifically into statue-idols. In the Solar Republic, "no statues are erected to anyone except after they have died. . . . Dead bodies are not buried, but they burn them to get rid of plague and to convert them into fire . . . and so as not to be suspected for idolatry." In fact, the only human images that are allowed are purely for the purposes of the technology of reproduction: "And these the pregnant women admire who apply themselves to the use of the race."[34] Transformational techniques are inextricably entwined with generating monsters, generating heroes, and generating gods. With each of these arts there is a danger of something going wrong, of some element of trickery and deception entering into the process.

No one is more aware of the thin membrane that separates good creation from diabolical illusion than Gaspar (or Caspar) Schott, a Jesuit pupil of Athanasius Kircher who worked closely with him in the Collegio Romano, editing some of Kircher's later works and putting his notes into publishable form. Schott's two great tomes on magic, *Magia universalis naturae et artis* (1657–59)[35] and *Physica curiosa* (1662),[36] can be considered the definitive, exhaustive statement on magic in the late *seicento*. There is nothing that one might want to know about these arts that is not contained in these copious volumes. What arts are these? "Chromatic magic, magic with mirrors, dioptric magic, telescopic magic, phonocamptic magic (about echoes), phonotectonic magic, phonurgic magic, phonoiatric magic, musical magic, symphoniurgic magic, thaumaturgic magic, static magic, hydrostatic magic, hydrotechnic magic, aerotechnic magic, magic arithmetic, geometric magic, cryptographic and cryptologic magic, pyrotechnic magic, magnetic magic, sympathetic and antipathetic magic, medical magic, divinatory magic, physiognomical magic, and chiromantic magic."[37]

Clearly, between the late sixteenth century when Porta was writing and the late seventeenth century when Schott was active, natural magic had become inseparable from its technological counterpart, artificial magic. Schott's definitions are designed to allow the practitioner to be able to distinguish between these licit arts and demonic magic, to which the first fifty pages of volume 1 are devoted. From Adam, through Seth, Enoch, and Ham, this primeval, divine wisdom survived the universal deluge. But, he warns, it is true that there is nothing so sacred, praiseworthy & pure that the fraud of the Demon and the malice of dishonest men have not ruined and corrupted; so that even the knowledge of hidden things handed down from Adam to his descendants and honored with the name of *Magic* was contaminated by the *Master Demon* with wicked superstitions and abominable teachings, so that the primeval and legitimate *Magic*, as soon as its name began to turn hateful for good men, was shortly so diffused like a plague in the world that no population has ever remained uncontaminated by it. From this corruption was born the division between *licit & illicit Magic.*[38]

The image of demon malice as a plague, as an uncontrollable contamination, flooding the world and corrupting all aspects of it, is a powerful, almost Gnostic vision of the spiritual forces active in the cosmos. One develops the picture of the magus, beleaguered in his laboratory, fighting valiantly to keep pure and uncorrupted the technical and transformational arts at his disposal. This vision is apparent in the definitions Schott provides. Magic is "the highest perfection of made and natural wisdom and hidden knowledge of things through which, with the wonderful application of the active with the passive, are made the common sense of man and everything on top of it."[39] Artificial magic is "the art or faculty of accomplishing something marvelous through human industry, using various instruments to that effect."[40] Schott's examples are already familiar to us: Archytas's wooden dove, Albert the Great's speaking head, as well as the mechanical statues that Kircher constructed for Queen Christina of Sweden and many other "*statuae automatae* in the form of humans and animals."[41]

As with Campanella, usefulness, industry, and instrumentality provide the conceptual ramparts against demonic invasion. Illicit magic is "the art or faculty through which, not by industry or artifice of man nor by the application of natural causes to certain things, but through the work of Demons, marvelous things are done that go beyond common deeds."[42] But how is one to know the difference between licit and illicit magics? Simple, says Schott, look at the means and ends that are employed. If the means and ends are good, then it is licit; if the magic employs human industry, it is good and licit; if it is for a wicked purpose or makes use of demonic works, then it is wicked and illicit.[43]

In Schott's vast compendium monsters appear in several contexts. Books 3 and 5 (*Mirabilibus Hominum* and *Mirabilibus Monstrorum*) of the *Physica curiosa*—consisting of some 230 descriptional pages—are devoted to teratology, carefully separated from *Mirabilibus Portentorum*, which occupies book 6. In the *Magia Universalis*, the monsters that appear are entirely transformational in character, popping out of instruments, images, and various marvelous recipes and contraptions. Many of the instructions are taken verbatim from Pseudo-Albert's book of secrets and Porta's *Magia naturale.*

Parastatic magic tells us how to make a man appear to have a head swarming with serpents, for instance: "Take the fat of a serpent, whichever you want, & a pinch of salt placed in that; and take a burial cloth, cut it in four parts, and then take the serpent fat, as described above, and put it on the pieces of the burial cloth, and make four bands out of it, & place them in four new lamps with sambuco oil, & place them in four corners of the house, and do as I have told you."[44]

The witches of *Macbeth* had no gruesomer nor exotic brews: "Fillet of a fenny snake, / In the cauldron boil and bake; / Eye of newt, and toe of frog, / Wool of bat, and tongue of dog, / Adder's fork, and blind-worm's sting, / Lizard's leg, and howlet's wing, / For a charm of powerful trouble, / Like a hell-broth boil and bubble" (4. 1. 12–19). The prose style of Pseudo-Albert's spells, reproduced (with added Latin translations) in Schott, has the same sing-song quality as Shakespeare's dramatic version. It is repetitive, written in the imperative mode, stringing together the same phrases over and over: "Pick up some fat . . . take some fat . . . put . . . put them . . . put them; Pick up some hairs . . . pick up the bone . . . stir it with the stick . . . put it above the door," and so on and so forth, as if the Magister Magus in instructing his pupils could enchant the reader with the very descriptions of his spells, creating a hypnotic rhythm with his words and inducing a state of submission to his commandments. The active, interventionist character of these texts is evident in the preponderance of imperative verb forms.

Yet Schott is first and foremost a "scientist." The formal, organizing principle of his text is a rigorous exposition of each discipline, divided into *Liber, Caput, Pars, Preludio, Syntagma, Articulus, Propositio,* and *Pragmatia.* Especially when it comes to optics, this peculiar combination of fact and fancy, rigor and hearsay is disorienting to a modern reader. In this *Magia Parastatica,* one might say, the original meaning of *poesis* (to make, and only by extension to fabulize) finds its most accurate embodiment: one constructs, fabricates, makes the machines in the interest of fantastic production. Science is harnessed to produce Protean, fantastic, and erotic images, which then provide the name for the technical activity: one of Kircher's many "new sciences" is dubbed "Proteus Sciathericus, sive Astrolabiographia figurata" for example.[45]

Magic, myth, and scientific operations were directly associated in these so-called catoptric experiments. As Jurgis Balstrusaitis reminds us, "Magic and mythology are directly associated to a scientific operation *par excel-*

lence, based on the incidental and reflected rays that transfigured and reversed the visible world." He further notes that "from time immemorial, catoptric equipment has offered both entertaining optical tricks and, at the same time, instruction on the laws and mechanisms of sight. This bizarre machinery, ever branded by a mystery, preserved its aura of prestidigitation and legend for a long time. The learned and the curious amassed them in their museums."[46] The association of monsters with parlor tricks and optical illusions that we find in natural magic manuals is carried over into museum collections of the period. Monsters appear in deforming mirrors, in catoptric machines, and in automatic theaters.

Metamorphosis in the Museum

At its height, the Musaeum Kircherianum housed in the Society of Jesus College in Rome must have been a splendid place to wander through. Along with Elias Ashmole's foundation in Oxford, it ranked as one of the first public museums.[47] Athanasius Kircher (1602–80), professor of mathematics, linguist, natural historian, mystic pilgrim dedicated to a synchretic unification of all the world's religions and philosophies, "that universal scribbler and rhapsodist,"[48] was a prodigiously prolific and insatiably curious intellectual. His wide-ranging eclecticism showed as much in his collection as in his massive, numerous monographs. Gabriel Clauder, physician to the duke of Saxony and member of the Academy of the Curious, relates that Kircher's incomparable brain had produced more works than warriors poured forth from the Trojan horse: "Not our Europe only but the whole world knows how much light he has shed on many sciences in this current age by his laborious dexterity and rare keenness of genius."[49]

Centrally situated at the heart of the global Jesuit network, receiving scholars, letters, and specimens from all parts of the world, Kircher was ideally suited to become curator of a collection of antiquities made over to the college in 1651 by Alfonso Donnino. "The museum," writes William Schupach, "though the property of the college and housed next to its library, became a monument to Kircher's diffuse intellectual interests." The collection of antiquities was quickly expanded with "*exotica* from Egypt, the Far East and the Americas as well as with mechanical, optical and acoustic instruments of his own devising."[50]

The museum was immortalized at its apex in a catalog published in 1678 by his assistant in mechanics, Giorgio de Sepi, and in the engraved title page we are given a sweeping view of the L-shaped hall, with the three rooms on the left that housed mathematical, hydraulic, and experimental machines from the mechanical arts, ancient manuscripts in Syr-

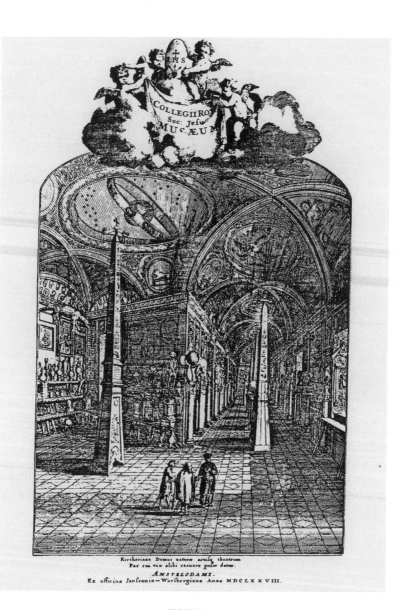

Kircheriana Domus naturæ artisq theatrum
Par cui vix alibi cernere posse datur.
AMSTELODAMI.
Ex officina Janssonio-Waesbergiana Anno MDCLXXVIII.

FIGURE 4
The vestibule of Kircher's museum, from
Romani Collegi Societatis Jesu musaeum celeberrimum, 1678.
(Reproduced from Adalgisa Lugli's *Naturalia et Mirabilia*
[Paris: Éditions Adam Biro, 1998], p. 117)

ian, Hebrew, Greek, and Chinese, along with illustrated books, and finally, in the third room, various automatons (see figure 4).[51] The perspective point in this engraving is almost vertiginous: the spectator floats from above, suspended at mid-point in the cavernous hall, as if hung from the ceiling like the stuffed crocodile one espies across to the left at almost the same height, appended from the vault, head pointing downward as if to swoop to the floor like some wingless dragon.

Kircher himself, costumed in his characteristic tricornered hat and dark robes, is engaged in conversation with several fine gentlemen who are cloaked and wigged with sword and walking stick in hand. As the elder aristocratic visitor, his back turned to us, gesticulates to Kircher, the Jesuit scientist gazes upward to our eye, as if monitoring our intrusion into this highly selective and careful ordering of artifacts that structure his historical, experimental, and ethnographic universe.

It is the spaciousness of this museum that strikes the viewer who is familiar with other academic collections of the period. The vast open space, the serenity and containment, the absolute grandeur are so different from the majority of other *studioli,* in which walls are generally lined and drawers stuffed to capacity, in frustration at the inadequacy of the collector's efforts to create a perfect microcosm, to re-create Noah's achievement of bringing together in one enclosed space at lease one specimen from all known species of flora and fauna. Manfredo Settala's museum, for instance (see figure 5), dating roughly from the same period

FIGURE 5
Manfred Settala's museum, from P. M. Terzago and P. F. Scarabelli's
Museo o galeria adunata dal sapere, 1666.
(Reproduced from Adalgisa Lugli's *Naturalia et Mirabilia*
[Paris: Éditions Adam Biro, 1998], p. 215)

(1666) sports not one magnificent, solitary flying crocodile, but five smaller ones: two at each end of the ceiling, and one hung vertically at the entrance, along with swordfish, marine monsters, flying fish, ostrich eggs, tusks, giant crustaceans, and cherubim who artfully hold up the drapery.

What "hold up" the ceiling in Kircher's museum, instead, are obelisks, punctuating the floor and vertical space in a linear series along the corridor, and one to the left of the angled hall. They stand like venerable trees, monuments to the cultural success of the Jesuit *imperium traslatio*, willing our gaze upward toward the tastefully painted, sparsely decorated vaults of the ceiling, one of which seems to open up to the heavens. The globe of the earth, ringed by a belt inscribed with the zodiacal symbols, rotates augustly at a forty-five-degree angle among the constellations of the fixed stars.

This museum creates an ethereal space of contemplation. Directly in front of our gaze, in the exact center of the engraving, behind the three living figures below him, stands a human skeleton, mounted on a pedestal in a casual pose, meeting our gaze with a nonchalant stare. Mortality and immortality, universal vastness and terrestrial particularity engage each other in a pleasing dialectical play in this sumptuous hall. The burial urns and sepulchral inscriptions and the skeleton that greet the visitors in the vestibule serve to ground them in human history and human time before they lift their eyes upward along the vertical axis of the obelisk, to the lofty heights of the eternal stars.

Mortality also intrudes in this museum through the ravages of time: "After Kircher's death in 1680 the museum was neglected: automatons fell into disrepair, organic materials rotted and objects displayed to the public without supervision were stolen. In 1698 Philippo Bonanni (1638–1725) was appointed curator."[52] Bonanni emerges out of this museological narrative as a heroic figure engaged to do battle with Time: in surveying the automaton collection he observes that "some of these sorts of objects were preserved in the museum and they fell into disrepair, forgotten out of negligence or because of the destructive action of time. However, they have been restored and many of no lesser value have been added to the others."[53]

What place have monsters in this erudite set of objects, we might wonder. This is no curiosity cabinet, no thrill-seeking *wunderkammern*: it is the exquisite fruit of an extraordinary mind, one nourished and developed to the speculative heights of its contemporary culture. By Bonanni's time, a few monstrous specimens are to be found collocated in the marine plant and animal section: an "ossa Gigantum [Giant's bone]," a "Canis monstrosa [monstrous Dog]," and some "Sceletha diversorum Animalium, & varii foetus humani [Skeletons of divers Animals, & various human fetuses]" appear on pages 280 to 282 of the catalog, but they form a minuscule part of the twelve classes of objects that make up the restored museum. The only place that monsters really flourish, in fact, is in class nine: "Instrumenta Mathematica." By now, this shifted discursive context should not come as a surprise. As much as descriptional monsters lose their prodigious and provocative qualities, as their familiar taxonomies begin to weary, transformational monsters acquire a marvelous aura about them. Let us examine more closely a few of these monster-making machines.

We are fortunate to have several sources at our disposal. Kircher himself in several textual instances describes some of his favorite transformational *machinemata*, or mechanical contraptions. The most complete version comes from his *Ars Magna Lucis et Umbrae* in part 3: "Magia Catoptrica, or on the prodigious display of things by use of mirrors" (841–907). After a theoretical discussion on the principles of spherical, conical, parabolic, and elliptical mirrors, an inquiry into whether Archimedes was really able to set fire to enemy ships from a great distance by focusing the sun's rays through a lens, instructions on how to construct a statue that will let forth a prodigious sound at the rising of the sun, an overview of the science of *parastasis* (which we glanced at in Schott: activities like projecting dragon images wherever you like, or using a "polylychnium" candelabrum to create multiheaded images at will), and other techniques useful for religious purposes such as projecting images of idols and making them float about in the air (the art of *technasma*), we come to the science of metamorphosis, "seu transformatione Catoptrica."

"Someone refers to having seen not long ago," Kircher begins,

a book attributed to John Trithemius in which the author avows the transformation of human beings into any animal; yet (says this person) nobody has been able to grasp the reasoning behind this assertion: most people, therefore, have maintained that it cannot be done in any way without recourse to diabolical arts. However this may be, I do not want to discuss the basis of this apparent metamorphosis. I only say that there are many things about the nature of things . . . that can be easily deduced from their effects, however, only by those who are knowledgeable about the mysteries of nature. I know that many impious things have been attributed to Trithemius that are so alien that they are suspect. We will demonstrate very clearly, God willing, in some way in our *Arte Combinatoria*, how, rather, nothing appears to be more in accordance with Nature. I believe that Trithemius's promise could be substantiated in two ways: either with Catoptric Art or with a somewhat more ancient application of things thanks to which a man thinks he has been transformed into any animal.[54]

This attention-grabbing introduction is followed by a marvelous series of nine metamorphic techniques for turning men into monsters through multiplications, substitutions, reversals, enlargements, shortenings, dilatations, and so on. *Metamorphosis I, II,* and *III* make use of the same elaborate, complicated machine that occupies a whole room, or rather, that forms an entire enclosed space into which the spectator-participant enters or peers (see figure 6). An octagonal rolling drum, with the images of a sun, an ass's head, a lion, and other animals painted on each face, is hidden from view. A mirror is suspended at an angle from the wall above the drum and attached to a cord and pulley. As the spectator gazes up to look at his image in the mirror, the drum is rotated, creating a series of superimpositions of heads onto the viewer's body.

Says Baltrusaitis regarding this machine: "The person first saw the solar disk, symbol of cosmic power, then the animals paraded after it alternating with the person's own face, which continually changed its appearance. We are right in the middle of metempsychosis."[55] Metempsychosis? Perhaps. But I think what makes this game so pleasurable is not the philosophical idea of passing into other beings after death; it is the transgressive pleasure of dabbling in questionably natural, and, more probably, diabolical arts. These figures could also be sculpted; Kircher suggests that glass eyes and hair glued on bas-reliefs, with the addition of moving jaw-parts, all serve to enhance their realism. He should know: "I myself have this machine which has rapt everyone in great admiration when they see instead of their natural face, the face of a wolf or a dog or another animal."[56]

Metamorphosis II is accomplished by darkening the room, and projecting light through a window opposite the mirror. The spectator is regaled with ghostly apparitions floating about the room. *Metamorphosis III* projects the figures onto cylinders or cones placed on the floor, walls, and ceilings. *Metamorphoses V to VIII* make use of much simpler apparatuses, various shapes of simple mirrors: take an elliptical segment of mirror in order to deform the face of a man in a thousand ways: "If you look at this mirror lengthwise it will give you back yourself with no brow and then with the appearance of ass's ears: nothing could be more deformed than the nostrils and the mouth; indeed, they are so sinuously formed that, especially if you show your teeth as if in laughter, they could look like rocky sea-coasts: then immediately afterwards it will give you double and triple heads. The monstrous variety of apparitions is difficult to express in words."[57]

FIGURE 6
Catoptric machine that metamorphoses humans into monstrous animals,
from Athanasius Kircher's *Ars Magna Lucis et Umbrae,* 1646,
book 10, part 3, p. 901, fig. 4. (Courtesy of the Bancroft Library)

In *Metamorphosis VI* one learns how "to transform monstrous human faces into various animals." Take a mirror and shape it cylindrically near the bottom, with protruding "tumors": if you look at yourself directly you will see your face change into a crane's head and your neck will become very long; if you look obliquely there will be a stream of water with a crag or a rhinoceros horn that protrudes out of your forehead. "If you want to show the face of a goat, the mirror will be convex in two planes with slightly undulating surfaces, and you will see yourself in the form of a hideous Satyr, with horns, wrinkles, and a ridiculous opening for the mouth. The mirror will even show the face to be ardent and red with the face of a drunkard if you put it under a red filter. If you form the mirror with surfaces that branch off you will see the head of a deer. In a word, no monster is as hideous as how you will see yourself formed in a mirror adorned in this fashion."[58]

The last technique—*Metamorphosis IX*—is something of a letdown. Certain substances, when ingested, will cause people to *believe* they have been turned into an animal:

Let it be known that there are some natural things which, soon after they have been ingested, by exerting their sway on the imaginative faculty transmute a man into the thing to which he is most inclined. . . . There are also other things which, if eaten as a food, transmute and transform people into Cats, Dogs, and Wolves in the way described. But since these things have been created beyond the limits of our art and because of the many evils that could come of them, they must not, nor cannot be made public. And we, venerating in them the greatness and admirable majesty of Nature, consecrate them to absolute and perpetual silence.[59]

Transformation is a hotly debated topic in demonology and natural magic texts, in which almost all authors assure their worried readers that no human or diabolical agencies are capable of transforming human beings into animals. Francesco Maria Guazzo devotes several chapters of his *Compendium Maleficarum* to reiterating this important point. The question is usually framed in terms of whether witches and demons have power over external objects, "whether witches can by their art create any living thing," and "whether witches can transmute bodies from one form to another." The answer hinges on the power of the imagination over bodies, or of mind over matter, and why it is that a certain material organization properly entails a particular spirit or soul:

No one can doubt but that all the arts and metamorphoses by which witches change men into beasts are deceptive illusions and opposed to all nature. I add that any one who holds the contrary opinion is in Anathema. . . . For a human soul cannot inform the body of a beast, any more than the soul of a lion can inhabit the body of a horse, or the soul of a horse the body of a man: because every substantial form, to be true to its own nature, requires the peculiarly adapted dispositions and physical organism which are natural to its own body, and the soul regulates the motions of the organic body. Therefore, as I have said, no animal's soul can inform the human body, and no human soul an animal's body. The belief in such monstrous transformation is nothing new, but was firmly held by the Ancients many ages ago. . . . But, as I have already said, no one must let himself think that a man can really be changed into a beast, or a beast into a real man; for these are magic portents and illusions, having the form but not the reality of those things which they present to our sight. For the devil, as I have said elsewhere, deceives our senses in various ways.[60]

We have understood that transformation is a diabolical affair, involving the displacement of spirit into an inappropriate body. What makes it diabolical or unnatural is that organic bodies are arranged in such a manner that *only* the human or animal spirit that normally inhabits them is appropriate. For any other spirit to inhabit that arranged matter is wicked, impious, anathema. Why then, would Athanasius Kircher, top intellectual of the Jesuit College at Rome, be delving into such questionably orthodox practices? The fact that he cuts off his investigation into techniques for human transformation with an avowal of their potential danger if fallen into the wrong hands and minds is evidence that he himself was aware of their demonic possibilities.

Baltrusaitis, in analyzing these machines, stresses the illusory nature of the monstrous images that they produce—"Machines and mirrors provoke the same wonders and the same illusions." He interprets the spectacle that they provide as a purely fictional *alter mundus,* a space of evasion and irreality "where the multiple images form and reform in regions that do not exist yet which the eye cannot deny, that join up with the Absolute."[61] The fact that Kircher's machines only create an *illusion* of transformation is even more damnable, though, since it is precisely through illusion of the senses and trickery of this sort that the devil operates.

In fact, illusion, mirrors, deception, and diabolic intervention are a common cluster of attributes to be found in seventeenth-century treatises on artificial magic.[62] What the modern reader like Baltrusaitis perceives as a contradictory intersection of "exact science and depraved, fairy visions" is not contradictory at all in the terms of the *seicento.* Deception, illusion, and exact science, harnessed in the form of technological experiments into transformation of the human form, all take part in the same cohesive, unified investigative project: *to test the possibilities for inducing changes in spirit by manipulating changes in the form of matter, and vice versa.*

Baltrusaitis insists on the contradictory relation between imagination, fantasy, and scientific research that these catoptric machines represent: "Catoptric collections were all conceived with these contradictions. The museums of Rome, Milan, Copenhagen are museums of phantasmagorias produced with precision instruments."[63] Yet we have seen to the contrary that the production of monsters, whether phantasmagoric, organic, or mechanical in nature, was a highly directed, internally coherent, philosophical and scientific activity. It was the fact that such productions threatened to transgress the "laws of nature" that lent them a ludic quality and not especially their promise for access to an evasive, otherworldly space of enchantment. They are entertaining because they run such a high risk, because they teeter on that fine edge between licit and illicit magics.

The reason why Athanasius Kircher, Gaspar Schott, Giorgio de Sepi, and their Jesuit colleagues at the Collegio in Rome were so heavily involved with these practices remains to be seen. The solution to this puzzle lies in another kind of monstrous machine that dominated European court gardens, theaters, and collections for over two centuries: that is, the automaton, a moving mechanical doll figured in human or animal shapes.

Authomata Instrumenta

Delightful, terrifying, disturbing, fascinating, shocking, demonic: these are some of the typical adjectives used to characterize automatons and their effect on human viewers. Pietro Scarabelli, in describing a "most beautiful statue of bronze" that walked across a garden, observed that "because of the stupor that such a motion occasions, whoever begins to observe it is rendered immobile."[64] Emanuele Tesauro, as we will see in the final chapter of this book, remarks that spectators of such marvelous machines are so enchanted by them that they themselves become frozen in astonishment. A passer-by is hard put to tell who is the petrified human and who the animated statue.

Their effect is potent; their affect is ambiguous. Terror and delight, so often associated with late Renaissance entertainments, are the watchwords for these particular mechanical monsters. One is tempted to see them, as Baltrusaitis did, as purely ludic entertainments, but Eugenio Battisti, in his chapter on garden automatons in the Renaissance, problematizes this possibility:

> They could seem, at times, to be a game: if the study of cybernetics had not demonstrated that it is precisely in automatons that man seeks "to reproduce some properties of the living substance in a material that is manageable and familiar, in the hope of finding a formula for life divorced from substance, which is generally the bearer of life," and if we did not have to observe that some of the most surprising attempts to surpass human limits have taken place precisely in this area [of research], like the creation of a perfect chess player or an invincible warrior . . . or in the search for a living yet eternal beauty that neither death may cut short nor time wither. The automaton, then, is in many aspects the means by which man projects himself beyond his existential limits, magnifies his forces, accomplishes in the concrete—and not just by pretending or describing—the marvelous.[65]

As Battisti points out, court-constructed automatons, placed in grottos, in parks and gardens, served as "a precious mediation between human and statuary choreography."[66] What distinguishes automatons from statues and humans is precisely their characteristic movement: it is neither continuous and smooth like human beings, nor still and immobile like stone and bronze.

Another important point that Battisti makes about automatons is that they boast a long and, in many respects, continuous history, from the third century B.C., to the sophisticated school at Alexandria, to the rediscovery of Vitruvius and Hero of Alexander's texts, leading to an explosion of mechanical engineering in the late Renaissance. Battisti also points out their ubiquitous connection with satanic imagery: "The relevant literary citations, both ancient and medieval, that maintain that automatons are invested with a benign or evil power are infinite. For the most part they are moved by a mixture of desire and terror at seeing the sacred image move by itself, transmuting from a manufactured object into an active principle: and for this purpose, in order to stimulate the emotions of the faithful, as far back as ancient times, sacred statues were constructed with moving parts; just as, in order to create terror, animated demonic masks appeared everywhere."[67]

Franco Lucentini, in a short but densely packed article entitled "Automatopoietica" provides one of the "infinite" anecdotes that Battisti refers to: "Albert the Great, between astrological studies and others on the formation of monsters, ventured—at least that's what they recount—to construct an android. Saint Thomas, horrified, destroyed his creation as a work of diabolic magic. And for two and a half centuries, the authority of the Angelic Doctor sufficed to ward off any self-moving perils."[68]

The fact that Albert the Great was previously engaged in astrological studies, immediately followed by teratological studies, when he turned to construct his android is quite fitting for our argument: as we have seen in Campanella and Porta, these activities would seem to go together quite naturally. Yet another tale of a living doll, a talking head this time, that Albert constructed while taking time off his studies of monstrous generation also finished badly: "thrown overboard by the sailors, in order to calm a sea-storm that they believed to have been caused by or because of [the automaton]."[69] The definitive judgment, we might say, on these troublesome creatures ("troublesome" for the way they tease definitions of human and nonhuman, bothersome for the paradoxical conceits that they generate, like Varchi's double-bodied specimen) dates from the fourteenth-century Nuremburg chronicles: "Rotating mechanisms that operate gestures and strange follies come directly from the demon."[70]

I think we can take as a given, then, that automatons have always been associated with the realm of the sacred. Lucentini attempts to show that in classical times automatons were viewed benignly, and only in more recent times have they been associated with horror. He explains this gradual passage from the classical image of constructed ideal beauty, peaceful and respectable companion of man, to the romantic monster or creature of horror, as due first of all to a Christian reviling of the flesh and, later, to the Cartesian substitution of mechanistic organs for flesh. Lucentini sees this process culminating in Swedenborg's anti-encyclopedic backlash in the "Maximum Man, formed of all the spirits that make up the universe and charged with routing forever the monsters of materialism."[71] But as we know, especially from a reading of René Girard's works on the scapegoat, an association with the sacred is always ambivalent: godly or satanic attributes float indefinitely around the sacred object and are fixed to it depending on the social context it occupies and the mood of the collective. "Godly" and "diabolical" are flip sides of the same coin.

This essentially ambivalent functioning of the sacred serves to clear up another seemingly paradoxical fact. In spite of the predominately satanic associations with automatons in Christian writers, the accounts we have from not only ancient Greece and Rome but also through the fall of Rome to medieval times suggest that these kinds of marvelous machines have always been used to induce sacred wonder. In fact, it is precisely in temples and churches that we would expect to find them.[72] This answers our query as to why Jesuit scientists were so active in producing these mechanical wonders.

We are now prepared to go back to Kircher's restored museum and to enter into one of the small rooms on the left of the corridor. Bonanni will be our tour guide once again. "*Authomata Instrumenta*," he explains, "are instruments composed through Art in such a way that they seem to move by their own will,"[73] a definition that stresses (1) they are unnatural or artificial objects and (2) their movement, the fact that they seem to move of their own volition (*sponte*), is their distinguishing characteristic.

The first machine we come to is a wind organ figured with a Cyclops and singing birds; the second one is also a kind of "Organum Pneumaticum." We listen politely, marveling at the sound effects and then pass on. "Another small machine placed on the table attracts no less amusement for the eyes than for the ears: it is a Monkey dressed in the uniform of a Drummer boy. When you lightly touch a small toothed iron key placed on its side, the little machine suddenly seems to go into a rage, and starts to beat the little drum with all its might, striking with great precision the rhythm of beats usually used to incite soldiers to battle. In the meantime, it rolls its head, moves its eyes all around and, opening its mouth, shows its teeth ready to bite."[74] It strikes the spectators with fear (*timorem*), says Bonanni, by its jeering (*cachinnos*).

This taunting monkey is only a prelude to even more amazing mechanical wonders. A perpetual motion machine provides a philosophical respite, and then "while the spectator stays to look at this, all of a sudden a

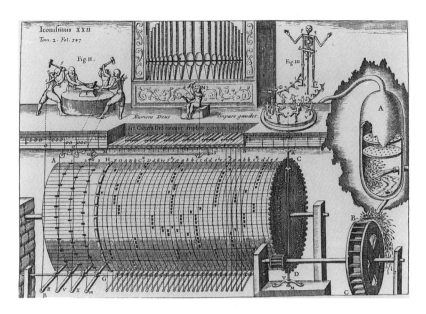

FIGURE 7
Automatic theater built according to Hero of Alexandria's descriptions
by Athanasius Kircher. From *Musurgia Universalis*, 1650, vol. 2, p. 347.
(Reproduced with permission from the 1970 reprint
by Georg Olms Verlag)

door opens wide and behind it sees a horrendous cave in which there is a hideous [*informe*] monster tied with chains which, as it tries to remove the snake around its neck and at the same time gives forth a terrible roar, the Dragon moves its gaping jaws closer in order to bite it."[75] Bonanni makes no pretensions to suggesting that any adult would be seriously taken in by such a spectacle. The affect it provokes is only mock terror, but children do find it truly frightening: "An object certainly terrible to children, entertaining instead to men, but not unworthy of the Mechanical Philosopher because of the artifice with which the movement and the voices are fabricated."[76] We will focus more closely on this use of entertaining machines for the purposes of scientific demonstrations in a later section. For now, another mechanical monster awaits our attention: "Not far from it lies a niche out of which the misshapen head of a Witch mocks the spectators by sticking her tongue out from her mouth with fiery bulging eyes and calling out in a terrible voice, to which a small wooden dog, struck by fear, responds with its own barking, so aptly performed that real dogs, excited by this, often add to these their own true barkings."[77]

There are many more such automatic theaters to be enjoyed in Kircher's museum (see figure 7), but I would like to pass on to a few other collections, just to show to what extent this conflation of monsters, demons, and machines is a widespread phenomenon in *seicento* collections. My account will be brief, but I hope sufficient to give an idea of just how pervasive this figural and thematic convergence is.

Manfredo Settala's museum, already mentioned in regard to the cluttered proliferation of objects hung from its roof, was one of the most famous in all of Europe, partially because of the widely disseminated catalog that made it accessible to those who were unable to visit it in person. (See figure 5.) The 1667 catalog is the third, expanded edition, translated into Italian by Pietro Francesco Scarabelli from the original Latin description by Paolo Maria Terzago. The preface boasts that "it is so famous in the World for its precious rarity, and for the variety and unusualness of the things that it contains" that its admirers have included queens and princes and dukes.

Monstrosity is well represented in Settala's museum in the form of organic specimens, catoptric mirrors, and automatons. In the section "On several rare Animals," Scarabelli draws an interesting parallel between Africa, traditionally coded as land of monsters because of its excessive heat and fecundity, and Settala's museum:

> Africa, because she is the fertile mother of monsters, merits having been dubbed a Theater of Marvels, which is why that ancient custom arose or was maintained: to curiously seek out *What new things has Africa bourne?* This itch will cease, in my opinion, for those who are pushed by the capricious, yes, but praiseworthy ambition to recognize even amongst the infinite species of the most monstrous and rare animals the excellent power of a God, if you will be so good as to admire the noteworthy quantity of animals that once were parts of monstrous Africa, or raised in most remote Woods; now immortal Cadavers, that are conserved in this Museum.[78]

Scarabelli's invitation to turn away from the "Theater of Marvels" that Africa provides to his own *museo* is an example of how the book, as a theater or gallery in itself, organizes and condenses into one small container, available to the panoptic view of the spectator, the unattainable, unassimilable, and overwhelming diversity furnished by the "real" world of nature.[79] The curiosity "itch" is a theme we will take up in the last chapter, but it is interesting to note the provenance of these testimonies to God's creative powers of generation: they take the form of monstrous African creatures and progeny of remote woods, as well as "immortal Cadavers": all faraway objects that are especially difficult to observe. The museum functions to conserve and collect together in one place; but it also serves to make visible what would otherwise remain inaccessible to everyday vision.[80]

This same idea of rendering visible the invisible monsters hidden in nature's vast storehouses is alluded to—with remarkably similar phrasing—in reference to another curious machine that often turns up in museums, generally collocated with the "mathematical instruments": I am referring to the microscope. Daniello Bartoli, in extolling the virtues of the *Ricreazione del Savio* (1659), comments on the marvelous powers of the microscope to bring into view the tiny bodies of insects, such "extraordinary and bizarre inventions of bodies, that human caprice, daydreaming, would never be able to imagine so many, nor such as them . . . and my interest will be, just by introducing your eye to a microscope, and up close to a multitude of these barely visible little beasts, to make you see in them things more marvelous than marvelous; so much so that you will confess until that moment to have been aware of only one tenth of what is beautiful and admirable in the world."

The microscope thus functions not only as a home museum but also as a surrogate world that surpasses the ordinary world to the extent that its magnifying powers enlarge it. The world observed under the lens constitutes both a synecdoche and supplement to the "real" world. Like Kircher's catoptric machines, the microscopic view transforms the ordinary into the monstrous, and through its technological creations pushes human imagination to its limits. Bartoli's conclusion is obvious: Why go to Africa, at so much bother and expense, when Africa can come to you? With the aid of a *microscopio* any studio can be transformed into a Libya: "If there is so much pleasure to be had in viewing monsters, and human curiosity rushes willingly to see them, there is no need to pass over the sea, and wander through the deserts of Libya, searching for them at your own excessive cost: every sod of earth is an Africa, in which many most strange and innocent monsters make their nests. . . . Have you heard described by the poets the Harpies, the Stymphalian birds, the Hippogryphs, and the Medusas, and the Furies, and I was about to say, the demons? You have some of all of these amongst these little animals, for which we thank God for not having made them with larger bodies or in a form that is visible to everyone."[81]

Manfredo Settala's "Africa" included crocodiles, armadillos, chameleons, hairy palm-sized testicles, a flying cat, a "previously unknown animal," as well as the head of a sea horse, a two-headed monstrous calf, and a siren's hand or "pesce Muscier." His collection also boasted a fine assortment of mummies.[82] Mirrors, lenses, telescopes and microscopes, mechanical clocks, "quasi" perpetual motion machines, and catoptric theaters occupy the first thirty-seven pages of the 330 total. Scarabelli quickly moves from a history of mirrors, their useful applications (from astrology, military engineering, to "public works and private Princely pleasures"), to a detailed inventory.

We are introduced to a trick mirror that "invites onlookers to a thousand jesting trickeries." By distorting one's body, one delights one's mind:

> With experience any curious person can satisfy his mind with many wondrous effects, standing at a proportionate distance from the Mirror, stretching out his hand with his arm, he will suddenly seem to see that an arm and a hand come forth out of the Mirror, and the members meet up, with this difference, that the stretched–out right hand appears, & is believed to be the left one. . . . Anyone who approaches, even having the height of a pygmy, will seem to have been made tall, as much as someone with a slender corporature will find himself made fat. Next, if he looks into the convex part, he will see himself extremely close up, as much as at a remote distance, without any harm, he sees himself quite tiny, and receiving Solar rays, they are reflected not by uniting them but by spreading them apart.[83]

Right made left, pygmy turned into giant, the world upside down are eternally present and accessible "without damage to oneself" simply by looking into one of these mounted, polished steel mirrors.

In another description of a dioptric machine, Scarabelli plays on the gap between appearance and reality: "A glass Mirror configured as an oblong quadrangle on the exterior flat surface, and on the interior scored into squares, by which the image of the object is broken up into each of the five areas, which shows the face of the person who looks into it wholly or partly represented with monstrous confusion, from which, along with the Latin epigram *Here are represented your many faces, multipled eyes, twin noses*, the saying *I abound with appearances alone* might be said to have come true for him."[84] Yet another polished concave surface is likened to a "Materno Utero [Maternal Womb]" from which ghostly apparitions seem to stop in midair.[85]

Another one "from its concavity enlarges the image, representing it at every small turn of the Mirror in various natural ways, yes, but disproportionately monstrous, which happens from the convex part, while the diminished image of the object transmutes into bizarre variations."[86] Scarabelli's commentary on this distortional mirror is revealing: while play, deception, and creative generation were foregrounded in the previous examples, this one provokes a meditation on the monstrous inconstancy of the human condition, translated into the comic and tragic modes. "From this Mirror Democritus would derive no less reason to laugh, than Heraclitus would to cry, together to laugh, and weep over the monstrous inconstancy of the human condition, about which Ovid would say: *Nothing is permanent in all the world. / All things are fluent; every image forms, / Wandering through change.*"[87]

The technological artifact (the deforming mirror), turned into a textual artifact (the catalog description), serves as an allegory (for the human condition) and pretext both for philosophical speculation (Democritus and Heraclitus) and poetic reproduction (the citation from Ovid). In this chain of production, image, artifact, technology, poetry, and philosophy peaceably cohabit one harmonious discursive realm. Contradiction, the kind that Baltrusaitis takes for granted, between exact science, high-precision technology, and poetic philosophy is obviously not a necessary relation, nor one that has always existed.

Even more striking in all these dioptric and catoptric machines—especially Kircher's metamorphosis room with the rolling drum—is the way that the spectator or the artificer *assumes the place of the monster*. Rather than creating specimens to put on display, these transformational techniques function by enhancing or manifesting the latent bestiality or monstrousness that now is represented as lying dormant inside us all. Monstrosity becomes pervasive, ubiquitous, always at hand and easily activated, rather than the mark of the unusual or the rare that arrives inexplicably from failures in nature's purposive effort, or sent by the gods. Technological instruments create monsters, transform innocuous beings into hideous ones, and discover real, invisible monsters where they are least (or most?) expected to reside: in drops of water, in human blood, in fleas and spiders, on the moon, in our companions' faces as they peer at us from the other side of a glass sphere, in our own faces as we gaze into a polished surface.

In some ways—ways we will explore more fully in chapter 4—monstrosity is interiorized during the *seicento*. Its locus was shifted from the field and the piazza to within the human heart. For example, just as Africa, the earth's monstrous womb, could be brought into one's own study, "Libyans" could be generated out of one's own body. If you could not go to Ethiopia, and did not have the means to purchase a microscope, perhaps the next mirror would have interested you: "A Mirror in which when one looks at oneself in it, even if you are of a very white complexion, you see reflected one's own image blackened in the guise of an Ethiopian."[88] Once again, we are struck by the conceits that these machines offer, the rhetorical archness of contradictory logics that they present: the spectator is the same and yet different; she is herself, and yet other: *candidissima* is opposed to *annerita*; self ("la propria imagine") is alienated in other ("un Ethiope"); and reality is here opposed to appearance ("a guisa di") through the ironic conjunction of *benchè*. Mirrors are supposed to reflect what is; they are expected to make visible the image of oneself that is otherwise inaccessible and reveal to the viewer what others habitually observe. These trick mirrors do the opposite: one does, indeed, see reflected "one's own image," but that image is no longer one's own; it is the image of someone else.

Amongst perpetual motion and other catoptric theaters representing a hunting scene, a naval battle, carriages with people climbing in and out of them, we come to "the head of a horrible Monster":

> A Pedestal, in whose upper, enclosed part one looks upon the head of a horrible Monster; with the simple touch of a trigger, lo and behold a door menacingly opens, from which issues such a monstrous Head with such a terrible roaring voice that it alone transports and fills with terror whoever hears it; From two little cannons that hang from both the ears by a thread, two Vipers shoot furiously out and, wriggling with a thousand twisted strands, provoke no little terror in the spectators; and while they consider the cause of their fear to have been in vain, and with joyful laughter return their spirits to their once-disturbed tranquillity, all of a sudden, opening out of a little window that one observes above the head, the spectators are bowled over by a new upset, when they are suddenly made to see an even more monstrous Head, while, unfurling a tongue from the opening between its lips and rolling its flaming eye between its frightening brows and moving its ears, which are Ass's ears, they are once again invited either to the terror of appearances or to the laughter of playful trickery.[89]

Scarabelli's (or Terzago's) commentary is always written from the point of view of the spectator/victim. It is not the machines themselves that interest him, as in the case of Bonanni, so much as the reaction that they provoke in the viewers. A little devil that one can submerge and surface at will in a vase of water—"artificio per verità dilettevole [a truly delightful artifice]"—provides the onlookers with "non ordinario diletto [no ordinary pleasure]." Another automaton in the form of a chained slave that screams and writhes "in addition to the ordinary confusion of his movements, in one instant both the eye and the ear of whoever watches it for any length of time are stupefied, with no little admiration on the part of the Spectators."[90] The movement of another machine provokes "extraordinary wonder in whoever observes it." For the "riguardanti" of the lenses that magnify their companion's face, "each person derives playful entertainment from the distortions of his companion."

"Delightful," "pleasingly deceptive," an invitation to the spectator's eye "to curiously submerge itself in a Sea of wonders": these are the typical effects and affects that these deforming mirrors and machines provoke.[91]

Except for the monster-automaton. As we read in Scarabelli's description, the spectators are never sure if they are safe, never quite convinced if the deception or the reality of appearances is more real. The monster automaton is the only one that seems to provoke any hint of genuine dismay, and its frightening aspects are precisely what make it the most shocking and impressive. For some reason the combination of the monstrous form and the technological matter has a peculiar power to disquiet the spectators. The question arises, then, and this will be the main focus of our next discussion: Why are automatons so often figured or shaped as monsters, and why are automatons considered monstrous in themselves?

The Origins of Idolatry

Lorenzo Legati's *Museo Cospiano,* dating from 1677, is in many ways indebted to Terzago's previously published catalog. Some passages are repeated verbatim and, in general, Legati is happy to emulate the work of his illustrious predecessor.[92] Many similarities exist, then, but also some important differences, both in the collections themselves and in their textual counterparts. One senses the presence of the very lively Bolognese scientific community in the endless collegial anecdotes that Legati narrates: his deference to the great natural scientist Ulisse Aldrovandi, who began the public botanical garden in Bologna and donated his own museum to the municipality; the frequent appearances of fellow doctors like Fortunio Liceti, Bartolomeo Ambrosini (editor of Aldrovandi's *History of Monsters*), and Ovidio Montalbani, whose *Cure Analitiche* (*sic*) is often cited. Organic monsters occupy more space and have more importance in this *Teatro di Natura,* and art is played off nature in a spirited rivalry, with Art coming out a clear winner, not only superior in beauty, nobility, and resourcefulness but even able to correct nature's involuntary mistakes.[93]

Every monster is an excuse to include some fellow academician's mediocre poetry or an equally inane Latin epigram. Conceits abound, like the monstrous baby girl who, although lacking a head, actually lived for a short time, giving rise to heated discussion among the eager scientific observers "so that the life of this Monster, however brief, was a long argument that vividly demonstrated that the Heart in animals is of greater necessity than the Brain."[94] If for the sake of a witticism a child's brief life can be likened to a long argument, we should not be surprised that the mother's life can be described as a fair exchange for "such a stupendous birth" as a two-hearted infant who died a few hours after being born, "following its Mother, who expired her soul in the act of delivering it, almost as if she could not bring into light such a stupendous Work, without compensating for the birth with a death."[95]

Nor should we be surprised to discover that two of the monstrous human specimens in Cospi's museum actually lived there on a permanent salary: the dwarf Sebastiano Biavati and his sister Angelica Biavati, "also a Dwarf, her very well-proportioned members being equally symmetrical to her height, which did not reach thirty Roman inches, even though she is 55 years old, living with her brother in service to Sig. Marchese Cospi."[96] Actually, at the time Legati writes, what remains of the dwarves are only their portraits. They are hung above the entrance to the museum rooms, each with the Greek inscription "Custode del Museo" (Custodian of the Museum).[97]

The catalog also reproduces an engraving of a double-bodied cat, preserved in a glass jar, giving rise to a cluster of epigrams, poems, and conceits. This generalized objectification of specimens for the sake of rhetorical and descriptional virtuosity reaches a height in the story of Sig. Montalbani's three-legged cat. In spite of its handicap, it was always a superb mouser: "It ran and jumped with incredible agility." Unfortunately, being endowed with a "ferocious nature, it always devoured its Offspring," and "later agitated one day by the furies of love, it was lost, nor did it allow itself to be seen in its usual house." Montalbani's reaction to the disappearance of his marvelous/monstrous cat? "Particular disgust," says Legati, since he had been looking forward to conserving its remains "just as Petrarch once did with those of his Cat, that even to our days are preserved"; upon this there follows a satiric Latin poem on Petrarch's dead cat by Antonio Querenghi "gentilissimo Poeta."[98] Taxonomy—the descriptive mode—as we have seen, evacuates the monster of any association with terror or horror; here the descriptive mode takes on the jovial character of "civil conversatione" among the male, educated, urbanite class of Bologna.[99]

We could spend much more time wandering through this museum and find support for many of the arguments and connections that I have made regarding Kircher's and Settala's museums. For example:

(1) Deforming mirrors are described as transforming the spectator into a monster, and confusing the human lineaments, "which often reproduces a Pygmy from a Giant, as in the Convex Mirrors; a Giant from a Pygmy, as in the semicircular Concave ones; or it makes a most monstrous face out of a very beautiful one, and sometimes a Chaos of extremely confused lines, representing anything but human features, as in several of the Concave Cylinders."[100]

(2) Legati's description of lenses and optical instruments are said to make visible the invisible: "Who will not say that it bestows wings on the human gaze, so that it may fly where it could not without them? and, almost as if to say, in order to see the invisible?"[101]

(3) Technology produces marvelous monstrosity from an ordinary object: a spider, when viewed under the microscope "is observed to be as despicable for the deformity of its snout, and for the horribleness of its whole body, as it is wonderful for the extraordinary number of its eyes."[102]

(4) The *microscopio* is also described as a complete, alternate world, a theater of nature in itself that encloses a world of objects: "And these are not outside of it, but inside of it, made into a receptacle and Theater both, of the most marvelous Works of Nature that are discovered inside it"—again, bringing into view "other parts that are invisible by other means."[103]

(5) Technological capacities are tainted with demonic associations; Legati defends their unusual powers with the epithet *innocente*. The microscope is said to work through "portentosa, ma innocente magìa [portentous but innocent magic]." Telescopes, too, are described as functioning "con innocente magia."[104]

But what interests me more at this point is to understand the connection between moving statues and monstrosity. Legati's archeological musings, I believe, point us in the right direction. Apart from a rag-tag assortment of mummy parts, the Cospi Museum also housed a fine collection of antiquities. These included sacrificial knives, vases and burial urns, sepulchral tombstones with inscriptions, ancient coins and medals, bas-reliefs, and finally, in book 5, a variety of religious statues.

Legati's essay "The Origins of Idolatry," with which we will close our reading of his catalog, contains a story that will sound familiar: "Saint Fulgentius, referring to the origin of idolatry among the Egyptians, would have it that someone named Sirofane, a rich man, driven by overwhelming love for a son who had passed away, in order to alleviate his grief had made for himself a statue of the deceased, and this came to be venerated by servile adulators by first adorning it with crowns of flowers and with offers of incense: and finally, like an almshouse to which everyone has recourse, it was recognized as divine."[105]

This narrative posits the beginnings of idolatry, that is, the creation of false gods in statues, in the same process by which it was believed in the seventeenth century that women could create their dream child by gazing at statue replicas. In both cases the absent, desired being is made present in matter by having a statue sculpted. In both cases, the yearned-for being is made animate through ardent desire: the son first through the father's "overwhelming love," and subsequently through the worshipers' faith which turns the statue into a (false) god. In the other case of the pregnant woman, the desired creature is called into being through the workings of the imagination in imposing the statue's form on the fetal matter. Both processes turn inanimate matter into (para-)animate beings through the workings of the spirit.

The similarities of the story gloss over more important differences, though. While the workings of the imagination on unborn fetuses could be harnessed and used to produce perfect children, they generally were blamed for deforming the female matter and causing all sorts of monstrous births. But no matter how disfigured, bestial, and unrecognizable the monstrous infant was, according to this explanation—accepted and propagated by scholars and medical men—there was absolutely nothing demonic about it. Imagination, or the desiring spirit, and its effect on female matter were considered to be absolutely *natural* phenomena. The same thing cannot be said for the unnatural practice of calling statues into life through the workings of the imagination.

The distinction lay in the difference between natural and artificial products, one that had been formulated some thousand years earlier by Aristotle, whose categories and definitions, as we have seen, remained firmly intact in early modern Italian thought. The difference between natural and artificial products, says Aristotle, lies in their source of motion and rest: "All things mentioned [the animals and their parts, and the plants and the simple bodies] present a feature in which they differ from things which are *not* constituted by nature. Each of them has *within itself* a principle of motion and of stationariness. On the other hand, a bed and a coat and anything else of that sort, qua receiving these designations—i.e. in so far as they are products of art—have no innate impulse to change."[106]

This distinction between natural and artificial things, based on their source of movement, is crucial to understanding why it is that moving statues were considered monstrous, portentous, sacred. What makes artificial products artificial is precisely the fact that "none of them has in itself the source of its own production."[107] To be more specific, in natural products (like fetuses) "the matter is there all along." Conversely, "in the products of art, however, we make the material with a view to the function," "and we use everything as if it was there for our sake."[108]

This idea, that artificial products are those that the efficient agent forms out of matter with a view to their use-value, is precisely what we would now call the technological attitude, the one adopted in producing artifacts like automatons.[109] Artificial products, by definition, "have no innate impulse to change": "*nature is a source or cause of being moved and of being at rest in that to which it belongs primarily.*"[110] So that for something to have had a soul, "it would have had to be a *natural* body or a particular kind, viz. one having *in itself* the power of setting itself in movement and arresting itself." Just as an eye is an eye only because it sees, "when seeing is removed the eye is no longer an eye, except in name—it is no more a real eye than the eye of a statue or of a painted figure."[111]

The point is that life or animation or soul can only and exclusively be said to inhabit a body that is self-organizing and natural. Automatons, statues that move, appear to make a mockery of all these definitions. That is why, in early modern Italy, when Aristotelian formulations of matter, form, and spirit were quite intact, moving statues could only be seen as somehow participating in demonic forces. And that is why so many prohibitions were attached to the use of figurines in natural magic and to animatelike machines in artificial magic. Matter that seems to move of its own accord could only mean trouble, transgression, unnatural conceptions.

This tricky distinction and the difficulty of heeding it when engaged in experimental technocrafting comes to light in Kircher's quest to construct the most lifelike speaking statue imaginable, one "breathing life from the mouth and from the moving eyes and from every part of its body . . . a statue that is perpetually prattling, sometimes producing human voices, sometimes animal voices, now laughing and snickering; now singing, then suddenly crying and wailing."[112]

We find the insatiably curious tinkerer at work in his *Musurgia Universalis*, not at what he calls "Musical Teratology" (book 9, part 3), which involves the production of prodigious sounds, but at the construction of *Instrumentis musicis automatis* (book 5, part 5). The challenge is this: "To build a statue both with and without moving parts which utters sounds." The Jesuit's introduction is characteristically mysterious and reflects the dangers involved in such a project:

> There are various comments on this wonderful *machinamenta* [machine, engine, artificial contrivance], and those who follow the teachings of the most secret philosophers believe without question that these can be constructed. Indeed they say that Albert the Great put together with admirable ingenuity the head of a man which could perfectly pronounce articulate sounds. As we showed in many ways in our *Oedipus Aegyptiacus*, the Egyptians had also built many statues which were able to pronounce anything at all in an articulate manner. Yet some people who repudiate these as contrary to the laws of nature can hardly be persuaded that such a machine could be constructed. They assert instead that the *machinamenta* of Albert and the Egyptians were either spurious or fake machines, or that they were architected by the work of demons in the same way as we

read that once the demons used to give answers in voice through oracles and statues. Nevertheless, many believe that such a statue can be produced, having been designed with such genius that it can pronounce some sounds in an articulate manner. . . . However this may be, we do not want to debate the famous head by Albert the Great or other *machinemata* by the Egyptians, but we do assert that we can indeed bring into being a prodigy of this kind, and without adding many words here we will teach how the statue is built.[113]

Kircher continuously stresses the effect that his handiwork will have on the audience. The whole point is to stun them with this *prodigium* (see figure 8), and that is exactly what these machines do to the spectators: "They will observe the movement of the eyes, they will admire the agility of the lips and the tongue, and they will look with stupor at this structure whose whole body breathes life; yet nobody will be able to understand with what artifice the statue has been built, or from what hidden machinery it has its motion."[114] Again and again Kircher refers to the secret, mysterious, hidden, abstruse *sacramenti* of this art, which the scientist will "penetrate." No wonder that there were some shocked disbelievers who doubted the purely natural basis of the automaton manufacturing process.

The desire to distinguish between demonic and man-made technology is an old concern, however, not one that developed solely in early modern Italy. In fact, demons, sacred powers, automatons, and mechanical wizardry have been associated in the West as long as written texts have existed. In book 18 of the *Iliad*, as Vulcan changes to meet Thetis at the door, he is escorted by his golden androids: "And in support of their master moved his attendants. / These are golden, and in appearance like living young women. / There is intelligence in their hearts, and there is speech in them / and strength, and from the immortal gods they have learned how to do things." Just before Vulcan was interrupted, he was busy fabricating a series of twenty *autòmatoi trìpodes* for the banquets of the gods: "And he had set golden wheels underneath the base of each one / so that of their own motion they could wheel into the immortal / gathering, and return to his house: a wonder to look at."[115]

Consider also this phantasmagoric passage describing the initiation of the emperor Julian ("the Apostate") in 361 A.D.: "Voices and noises, calls, stirring music, heady perfumes, doors that opened all by themselves, luminous fountains, moving shadows, mist, sooty smells and vapors, statues that seemed to come to life, looking at the prince now in an affectionate, now in a threatening manner, but finally they smiled at him and became flamboyant, surrounded by rays."[116]

FIGURE 8
Speaking statues designed by Athanasius Kircher. From *Musurgia
Universalis*, 1650, vol. 2, p. 303. (Reproduced with permission from the
1970 reprint by Georg Olms Verlag)

The problem, as we have seen in many texts, was how to distinguish
between *mêchanêmata* and unnatural practices. This spell from the Magical Papyrus in Leiden is typical of prohibited practices that link the art of
drawing spirits into inappropriate vessels with the diabolical art of necromancy: "How to awaken a dead body. 'I conjure you, spirit walking in the
air; come in, fill him with your breath, give him strength, wake up this
body through the power of the eternal god, and let him walk around at
this place. For it is I who perform this operation through the power of
Thayth, the holy god.' Say the Name."[117] Like many magical spells, it
reads like a parody of a biblical text. The magician usurps the place of God
in Genesis 2.6–7: "Then the Lord God formed man of dust from the
ground, and breathed into his nostrils the breath of life; and man became
a living being." Let us note that these arts "binding" demons into specially formed matter such as talismans and amulets are precisely what
Ficino, Agrippa, Porta, and Campanella were engaged in as they manipulated the sympathies and antipathies, the universal consensus of the
cosmos, to their ends.

The orthodox believer of the seventeenth century, however, was counseled that the soul "cannot move an external body." "When the soul of the
child killed by Simon Magus later appeared to change bodies according to

his magic, in reality it was a devil that operated, pretending to be the soul of the child," warns Campanella. "It is necessary to proceed cautiously in order to not attribute demonic effects to nature and lapse into error."[118]

Manpower/Machine-power

"Even for the pagans it is obvious that lifeless statues are not gods," Augustine once observed. "But let us [now] consider this question: How should one look at the powers lurking in statues? Can one have a pleasant relationship with them? Are they good and truly divine or the very opposite of all this?"[119] We have understood why automatons were considered monstrous and retained an aura of the sacred. But the question still remains: why were technological artifacts so often figured as monsters in the *seicento* and why did they continue to be represented as such?

Robert S. Brumbaugh shows convincingly that mechanics and mechanical gadgetry were first and foremost stimulated by philosophical debates, and not because of applied science or for the purpose of providing labor-saving inventions. Machines were modeled in ancient Greece to prove or disprove various theories about the cosmos, psychology, and natural forces. Automatons, for instance, answered questions about *psyche* ("soul") as both the vital principle that gave things life and the inner power of self-motion.[120] Brumbaugh provides a wealth of detail and fascinating illustrations of the interplay between mechanics, natural philosophy, religion, and social organization in the ancient world, yet he remains puzzled by the nature of this interplay. The fact that technics were not conceived as labor-saving devices, the fact that amusement or philosophical speculation were the "mothers of invention" rather than necessity remains a mystery to him.[121]

Langdon Winner, writing *Autonomous Technology: Technics-out-of-Control as a Theme in Political Thought,* sees that, on the contrary, since science and technology are concerned with the possibilities of control over natural forces "both politics and technics have as their central focus the sources and exercise of power."[122] Specifically in the West, one model of scientific and political power dominates: that of absolute mastery. "Much of the existing literature, for example, holds that technology and the human slave are exact equivalents, even to the point that they are functionally interchangeable."[123] This virtual equivalence between slave and machine is seminal for critical literature on the history of science and technology. It is also a key notion in Aristotle's *Politics.*

According to the ancient philosopher, the state is composed of households founded on the two relations of male and female, and of master and slave. The household exists to satisfy man's daily needs. Now in the property of the household, the master disposes of two kinds of instruments: animate and inanimate. "And so, in the arrangement of the family, a slave is a living possession . . . and the servant is himself an instrument which takes precedence [or comes prior to] all other instruments."[124] This idea of animate and inanimate instruments tightly links the social order with a corresponding model of matter, form, and spirit needed to naturalize or normalize institutionalized hierarchy.

In spite of what many say, slavery is not contrary to nature, insists Aristotle, because the distinction between master and slave is not simply a convention: "It originates in the constitution of the universe; even in things which have no life there is a ruling principle."[125] Master and slave are mutually necessary to each other's existence. In fact, the only situation in which one could imagine that "chief workmen would not want servants, nor masters slaves" would be "if every instrument could accomplish its own work, obeying or anticipating the will of others, like the statues of Daedalus, or the tripods of Hephaestus, which, says the poet, 'of their own accord entered the assembly of the Gods'; [or] if, in like manner, the shuttle would weave and the plectrum touch the lyre without a hand to guide them. . . ."[126]

The threat that automatons pose could hardly be clearer: if matter moves of its own accord, it presents a threat of breakdown, of collapsed boundaries, not only in the realm of natural forces—hence the association with demons and necromancy—but especially as a figure for the stability of the social order. Everything in the universe exhibits a distinction between "the ruling and the subject element. . . . Such a duality exists in living creatures as well: in self-organizing, natural bodies which contain their own source of motion and reproduction, the soul always rules the body."[127] If matter began to rule spirit, if it broke free and began to exhibit autonomous capacities like an automatic shuttle, racing across the loom on its own accord, the same horror would be unleashed that is created when body rules soul, or when the resistance of female matter defies the formative virtue of male semen in the womb, or when spirit is called into inappropriately formed matter. What would happen is that monsters would be generated.

The horror that is unleashed, as we have seen in Campanella and other early modern natural philosophers, is one of sinful deficiency, nature's impotence, lack, nonbeing, disorder, and chaos: in a word, monstrosity. Monstrosity is always and quintessentially this basic perversion in the "natural" order, whether in the biological realm, when matter prevails over spirit, stymieing its formative efforts; in the philosophical realm when nature is seen to err or be deficient; or in the social realm when the inevitability of the master-slave relation is questioned.

Conclusion

"For the fact that man was the maker of his gods did not mean that he was not possessed by what he had made, for by worshipping them he was drawn into fellowship with them."[128] The same thing could be said about man's machines. But this is not the place for me to pursue the monstrous machine theme into our present-day relations with technologies. It has been enough for me to show that sacred monsters did not disappear. What happened instead is that sacred associations left the biologically monstrous infants that were interpreted as portentous messengers from God and were displaced instead to mechanical monsters. Prodigy literature did die out, but that does not imply that prodigies ceased to exist.

There is one more connection I would like to bring to light, though. In Aristotle's metaphor, there is an implication that "something must be enslaved in order that something else may win emancipation."[129] This understanding of the master-slave relation of humans to technology subscribes to what we could call a logic of reciprocity. That is, humans hope to be liberated from labor through the perfection of industrial technology, but in order to do so they must give up something else. The question is, what is that something else? The highest form of technology is, of course, an autonomous artifact, self-regulating and self-commanding. But if automatons achieve such autonomy, from whom or what are they freed?

Some would answer: the human will. We relinquish control over our contrivances in order to allow them greater industrial powers, thus creating artifacts of increasing autonomy requiring less and less human intervention. As technology becomes more autonomous, humans assume a diminishing role in controlling and directing their operations. Machines break free of our will; in return, we are liberated from supervising them. The machine gains autonomy as humans relinquish it, but, at the same time, we become more dependent on them. The terms of the reciprocity become clear.

Fear and loathing of the machine are based on what some find to be a ridiculous idea: "that somehow machines will develop a volition of their own, independent of their makers and come eventually to change roles with man and make men their servants."[130] But is this such a ludicrous idea after all? Given the founding instrumental analogy between inanimate machines and animate slaves, the fear would seem to be quite logical. When the machine fails to perform, the servant is being disobedient, we seem to think. Fear of the oppressed is an essential tension in keeping such a dialectical relation intact. Winner attributes technological animism to the belief that there is a law of preservation of life at work, much like the law of the preservation of energy. "Where does this strange life in the

94 THE MONSTER IN THE MACHINE

apparatus come from?" he asks. "What is its real origin? the answer is clear: it is human life transferred into artifice. Men export their own vital powers—the ability to move, to experience, to work, and to think—into the devices of their making. They then experience this life as something removed and alien, something that comes back at them from another direction."[131]

This belief serves to explain why the products of our labor represent our highest fulfillment according to some political thinkers, and at the same time, our most draining and dehumanizing forces. For Karl Marx, when technology is part of a free, conscious, productive activity, it allows for the highest achievements of civilized man: "The practical construction of an objective world, the manipulation of inorganic nature, is the confirmation of man as a species being." "Productive life is . . . species life. It is life creating life."[132] When the means of production are wrested from the laborer, however, according to Marx, the resulting alienation turns the machine into a veritable monster: "An organised system of machines, to which motion is communicated by the transmitting mechanism from a central automaton, is the most developed form of production machinery. Here we have, in the place of the isolated machine, a mechanical monster whose body fills whole factories, and whose demon power, at first veiled under the slow and measured motions of his giant limbs, at length breaks out into the fast and furious whirl of his countless working organs."[133]

Winner is clearly justified in his conviction that "autonomous technology is ultimately nothing more or less than the question of human autonomy held up to a different light."[134] What remains misunderstood, however, is the terms of exchange that we referred to earlier. Reciprocity dictates that humans can only get by giving up. The most pervasive fear in regard to technology is that it will become autonomous and take our place, that the servant will become the master.

Science fiction literature and films abound with disturbing combinations of humans and machines in the form of androids, cybernetic creations, bionic men and women. No wonder: our medical reliance on prostheses, pace makers, and other electronic implants, biotechnology, and genetic engineering is not just a fear or a trope but a lived experience. In a very real way, our human bodies are becoming more machinelike, just as our favorite machines—like our "personal" computers that speak—take on a comforting anthropomorphic cast. Jacques Ellul makes the most pertinent observation, in my opinion, regarding the terms of the exchange: "Technique is entirely anthropomorphic because human beings have become thoroughly technomorphic."[135]

We fear, precisely, that a bargain has already been struck, that in exchange for total mastery of nature, we have already given up our human autonomy. For Liceti, Campanella, and their seventeenth-century colleagues, monsters were produced by the *disobedience* of matter. Now our fear of monstrosity stems from our too perfect *mastery* of nature. Not that the demons or the dead will reanimate statues or corpses, but that we will be summoned to take the place of the demon, that we will be indistinguishable from matter. The Faustian terms are nothing less than the annihilation of human distinctiveness, human spirit, or human essence, in exchange for mastery of nature and freedom from labor.

This is not a new fear. The Frankenstein myth is not just a product of the nineteenth century, nor is it only about human relations with technology. Its roots lie far back in the origins of idolatry and necromantic practices of calling up spirits to inhabit the lifeless bodies of the dead. It addresses the fear provoked by self-animated (or man-animated) matter that moves of its own accord. It also addresses our confusion at understanding what constitutes a human being: are we flesh or are we spirit? Are we body or are we soul? And what is the relation between the two? It would seem that the human condition is fraught with dangers: both to lapse into pure "body-hood" and to become indistinguishable from mechanically animated matter are pervasive threats. Either option is equally possible at any moment. In both cases, it is the monster that is called upon to signify that threat.

FOUR

Medicine and the Mechanized Body

We have seen that sacred monsters did not disappear due to a secularizing influence on popular culture; instead, they transmuted into mechanical artifacts. This is because according to ancient beliefs about matter and form that which is inanimate yet moves of its own accord was believed to be contrary to nature. The examples we have seen from Aristotle and from natural magic show that automatons, along with statues and corpses into which demons have been called, share this common monstrous attribute. In their most innocuous form, as entertainment devices, such unnatural bodies provoked wonder, whether they were found on museum shelves, toured about in the town square, or stood in gardens to titillate and astonish spectators.

The automaton, the moving statue, is monstrous because it appears to violate the natural order. A human being who creates a living apparatus competes with God, provoking fears of unholy creation, of diabolic intervention. The automaton indicates limits to human creation that must not be surpassed, especially when our creative power is used to replace human labor by technological industrialization. This fear already existed in Aristotle's conviction that a loom that weaves of its own accord would be as contrary to nature as a household in which slaves took to commanding their masters. Matter that moves of its own accord constitutes an unthinkable breach of hierarchy in the natural and social orders, potential threats to civic and cosmic harmony, and a deformation of formative power that could result in violent discord. These are the same threats posed by all monstrosities.

Conversely, harmony obtains in both the macro- and microcosmic worlds according to the precepts of *natural justice*, expressed in well-proportioned relations between social and natural elements. Justice, the quality of being just, is the early modern antonym for monstrosity. Like monstrosity, "justice" obtains in the physical, juridical, and moral realms.

For Plato and for Neo-Platonists, justice in the physical, metaphysical, and political realms are perfectly corresponding images: a well-proportioned body indicates beauty, identical with truth and goodness.[1] Furthermore, the body of the individual man is in every way comparable to the body of the polis.[2] In the same way that harmony is a product of balance between the rational, animal, and vegetative souls in the individual, the ideal state would maintain a peaceful balance between three classes of citizens.[3] As Plato explains in the *Republic*, "to produce health is to establish the elements in a body in the natural relation of dominating and being dominated by one another, while to cause disease is to bring it about that one rules or is ruled by the other contrary to nature."

Natural justice is a state of harmony in the body politic just as health indicates goodness in the body physical. Justice in these terms is a matter of the right elements of the soul "controlling and being controlled by one another, while injustice is to cause the one to rule or be ruled by the other contrary to nature." Virtue, it follows, is equivalent to "health and beauty and good condition of the soul," while "vice would be disease, ugliness and weakness."[4] Disease indicates latent deformations of character, the implicit presence of evil.

The same connection between moral deformation and bodily disproportion is crucial to Christian thought as well. The creation myth described in the book of Genesis represents Eve as the female principle (matter), called by appetite and irrational desires to taste the fruit of knowledge. Biting into the apple, the Ur-woman satisfies both her senses and her curiosity. Once Adam (the male principle of spirit or mind) tastes of the apple of knowledge he is expelled from the garden of earthly paradise for transgressing His Father's Law.

This story also depicts curiosity as a dangerous desire, since knowledge turns out to be, ultimately, knowledge of self. Self-consciousness and shame are what led Adam to fall from godlike immortality into a temporal world of decay and corruption, motion and change, suffering and extinction. Henceforth, man may yearn to draw himself upward toward angelic perfection, but he shall remain dragged down in his animal flesh, cursed by bestial cravings.

Furthermore, according to the doctrine of original sin, human beings arrive out of the womb sinners, fallen and corrupt by virtue of inhabiting the sublunar world of time, matter, and change. The just, those who are able to legislate a natural justice in themselves between the forces of reason and appetite, look forward to the day of final judgment when they will return to their God, restored to a peaceful relation between body and spirit.

The "civil war" that Christians must endure during their sojourn in the earthly body finds resolution only in the afterworld. Meanwhile, during the time of waiting represented by the course of human history, lost Edenic harmony between male and female, mind and body, spirit and matter is replaced by perpetual conflict between opposing physical and moral forces. Redemption can be obtained only through an arduous process of self-abnegation and control. This struggle is represented as a battle between the animal versus the human within, a dichotomy that finds expression in images of monstrousness.

The animal within that must be tamed in order to achieve peaceful harmony in both the individual body and the collective body politic is an integral concept both in medical practice and in political theory in the early modern period. It stands at the base of the revived, ancient science of physiognomy; it also informs disputes between supporters of Galenistic medicine, opposed to followers of the new mechanistic and chemical schools. In Galenic medicine disease is conceived as disproportion between the four humors (blood, phlegm, black bile, yellow bile), while health is achieved through their proper tempering.

The beast inside also figures prominently in theories of social contract elaborated by juristconsults as early as the sixteenth century and as late as the eighteenth by the French *philosophes*. In natural law theory, the animal within appears specifically in the troublesome question of sociability, that is, whether humans are naturally sociable or whether they require an external state apparatus in order to legislate and enforce a commonwealth. Not by chance, both the prince and the state are often figured as monsters. In the case of Machiavelli, the prince is counseled to think of himself as a wolf-lion or a centaur; in the case of Hobbes, the state is conceived as a Leviathan and figured according to the mechanical workings of an automaton.

The beast that lies within is a commonplace that serves to define the parameters of both physical and mental health. In *Madness and Civilization*, Michel Foucault paints in broad strokes two epistemic shifts regarding madness in terms of the beast within or without—first, from the Middle Ages to the Renaissance and, second, from the Renaissance to the Classical period. Leaving aside evident questions regarding use of periodization, we could sum up his conclusions thus: in the Renaissance the animal steps out of the world of legend and moral illustration, figured as the values of humanity in medieval iconography, to reveal the secret nature of man.[5] Since madmen are morally defective, victims of their own delusions, a catalog of the insane—the misers, slanderers, drunkards—reads as a catalog of the vices and virtues. "It thus gives access to a completely moral universe."[6] During the subsequent age of confinement, the bestiality within is exteriorized and put on display: "Madness had become a thing to look at: no longer a monster inside oneself, but an animal with strange mechanisms, a bestiality from which man had long since been suppressed."[7] Foucault accordingly sees animal metamorphosis in the Classical period (i.e., the seventeenth century) as a secularized, zero-degree sign of human nature in a state of nature, that is, our "true" nature.[8]

Foucault is right in stating that a history of madness is in some fundamental way tied in with a history of bestiality, identified as the essence of monstrosity, but we need to look more closely at how and when these shifts occurred. One of the aims of this brief survey is, accordingly, to explore in detail such concepts as "within" and "without," "moral" and "physical." At all times, then as now, the great Platonic metaphor which denounced the spirit's corruption in the folly of bodily sin, as Foucault describes it, not only holds sway, it founds our postmodern imaginings about our relations with our bodies, whether those imaginings are based on man the machine, man the rational animal, or man the impossible hybrid of matter and consciousness.

As we shall see, whether one considered oneself a follower of the ancient Hippocratic and Galenic traditions or of the new chemical and iatromechanical schools was irrelevant to a core belief in the power of the soul or of the passions to influence one's bodily health, and vice versa. This did not change with the age of mechanistic physiology. What remained to be discovered and defined was not a basic correlation between body and soul but only the exact manner in which immaterial forces, including animal spirits, magnetism, and gravity, were able to affect material bodies. Consequently, at the heart of these imaginings of old and new sciences lies an intense preoccupation with discovering the nexus between matter and consciousness.

Man the Microcosm

"Everyone knows that amongst the Philosophers it is a commonplace that the monster in the body is a monster in the soul, and being a monster in the soul, what can be expected of such a person, what should become of him, if not evils and misfortunes."[9] So reasons Porta in his work on human physiognomy. One would expect that this sixteenth-century work represented an old cosmology, that of man the microcosm, and as such was destined to lack credibility with the changing tides of mechanism that swept over the seventeenth century. However, the success of Porta's version of physiognomy, based partly on the apocryphal authority of Aristotle, belies this view: Thorndike reports twenty-one editions of the work (first published in 1586) in Italy by 1655.[10] It does not take much intellectual squinting to see the modern "science" of phrenology, for example, and the unabated use of palm reading as continuations of the physiognomy tradition. Indeed, many of our own contemporary psycho-therapeutic techniques are based on the notion that psychic trauma is stored in the body and can be read according to a symptomology that links past experience with present somatic pain. For us, too, it is a "common-place" belief that what the body displays in its proportions and dispro-portions intimately and directly echoes the harmony or disharmonies of the soul.

The basis of physiognomy is an ancient one. The authorities oft cited by early modern treatments include Plato, Socrates, the *Physiognomy* by pseudo-Aristotle, Polemon's *Fisionomia*,[11] Galen, Seneca, and a host of Arab and Christian writers. The origins of the science, as narrated by Porta and echoed in every treatise, is a functional one especially useful to princes; that is, to discern the true character of people beyond their possi-bly deceitful appearances. Porta expresses this succinctly:

Sometimes one discerns beneath the appearance of a benign man, as Seneca affirms, the soul of a wild beast, on the contrary, more savage than the most cruel wild beasts. For this reason, in order that men should not be able to deceive others, Socrates ardently desired that there be a window in the chest: so that in this way there could not be hidden a duplicitous heart, but that it would be possible for each person to discover desires, thoughts, truths, and lies. Such a just wish on the part of Socrates due to such a great evil is fully satisfied by Physiognomy, praised, researched, discussed, and held in the great-est veneration by the most praised writers.[12]

The ability to discern and identify latent animal tendencies, or as Soc-rates put it, to open "a window in the chest," is seen as essential for the development of good citizens, not only to be able to positively encourage children in their future pursuits, but also to eliminate those that are tainted from birth. Echoing Campanella's desire for eugenic control in the City of the Sun, Porta sees physiognomy as a primary instrument for control in "well-ordered Cities" and an instrument for encouraging civil harmony.[13] He describes in praiseworthy terms how "the Spartans were accustomed to not allow the father and mother to nurse their child, but they must first bring it to a tribunal where the wisest judges sat." If the infant body displayed signs of robustness and was well-proportioned, and was thus able to perform its soldierly duties in defense of the republic, it was returned to its mother to be nursed. But if it manifestly lacked the necessary "wit and good judgment" to govern the republic, the tribunal "carried the baby to Mount Taygetus and casted it down from the summit . . . judging unworthy of life those who were not fit to be useful to themselves, or to their country."[14]

What makes possible this sort of prescient diagnosis is the fact that Man is a little world, made in the image of God. Just as correspondences exist between microcosm and macrocosm, that is, between the human body and God's wider creation, there are correspondences between the physical and the spiritual aspects of the human body. Cornelio Ghirardelli, whose work we will discuss later in detail, expresses this concept in terms of the "great World machine." If this great machine has prompted so many thinkers to probe its workings, he reasons, "so much the more this little world of Man, in which it seems that the whole is concentrated (as in a compendium of perfections and prerogatives)."[15]

PHYSIOGNOMY

For an overview of the principles of this science let us turn to Francesco Stelluti's *Della Fisonomia di tutto il corpo humano.* Stelluti, a member of the Academy of Lynxes, is credited with publishing the first systematic observations with the microscope, on honey bees, in 1625.[16] That he saw fit some twelve years later to provide a comprehensive, taxonomic work on physiognomy in the form of synoptic tables is further evidence of the naturalistic basis of this discipline. Stelluti, whose aim is to present "what the Greeks, Latins, and Arabs have written about this matter" provides 155 pages of correspondences in the form of flow charts, presented horizontally across each page.

He begins by locating physiognomy within the broader category of divinatory sciences. Those based on art include necromancy, bird augury, and reading of dust, water, animal intestines, and so forth. Those based on nature include melancholy ("whose force and violence separating the soul from the body often causes prediction of other things"), sineopa ("when a man, as if dead, whose spirits are recalled from the office of his members through the help of his heart, and then returned to himself, predicts future things"), and dreams. Physiognomy, the fourth category, is presented as a medical art, derived from Galen and "founded on natural principles. . . . [It] is most trustworthy [*verdadiera*] and useful, because knowing the inclinations and vices of others by means of it, we may see and medicate our own."[17]

The next chart shows reciprocal correspondences between soul and body. Stelluti explains that "there is a great confederation and brotherhood and correspondence" between our corporeal and incorporeal parts. He bases his observation on firsthand knowledge: "Our experience shows that the soul always suffers with the suffering of the body, & the body is travailed and afflicted at all times, that the soul experiences considerable passions; and that medicating the body, the sick soul becomes healthy, as it happens in madness, an illness of the intellect."[18]

The chart that follows shows the signs in animals that indicate their usefulness to man. A hunting dog to be avoided, for instance, has a tall, small, disproportioned body with a skewed look to his eyes; he is deformed, ugly, with twisted feet, lazy, and melancholy.[19] Next we pass to human habits, readable from the temperament (the body, the heart, the brain); the Galenic humors (blood; black bile, indicating melancholic humor; yellow bile, indicating choleric humor; phlegmatic humor); and finally a series of indicators that provide the structure for the rest of Stelluti's charts. This list provides an overview of the physiognomic approach: the inclinations of man can be known from parts of animals; positions of the heavens; various qualities of the locale; apparent habits; the stars; foods; the person who nursed him; age; opposing signs; similar passions; parts of man, of woman, and their habits; the city-dweller; the savage.

Looking up the person whose brain is composed of a mixture of hot and dry humors, for example, we are told that he will exhibit the following characteristics. His hair grows immediately strong, kinky, and black then turns to red as a youth but quickly falls out, leaving the bald head hot to the touch. He sleeps little, but deeply, and has agile control of his animal functions. He is a quick thinker with acute senses, hardly bothered by the cold, and while in his youth he enjoys all his senses, in old age they will fail him.[20] Such a person, according to Stelluti's charts, has a thin, dry body, covered in dirty blond hair. His nature is impetuous, irate, and bold. He eats a great deal and is ambitious for honors, and being endowed with a great soul, he is more often prodigious than generous, being audacious, strong, deceitful, and astute.[21]

The next section lays out the exact characteristics of each animal and its habits as an aid to interpreting the habits of humans. It reads like a Chinese zodiac. The pig, for example, who is somewhat bestial, absentminded, irate, proud, lustful, incapable of any art, dirty, and insatiable, exhibits the following physical characteristics: straight, coarse, thick hair; flat forehead; small, deep-set eyes with eyebrows inclined toward the nose and lowering toward the temples, and a big nose with a forward jutting mouth and a very thick neck.[22]

Book 3 establishes the connection between moral virtues or vices according to signs of character (just, evil, fallacious, thieving, and so forth) and corresponding physical signs, and relates each moral type with appropriate animals, including woman as a species of lesser animals. For example, a just man has a well-proportioned body, with dark hair, prominent, wide eyes, a deep voice and is related to the lion and the elephant.[23] Not surprisingly, a bad man has an ugly face with long, close-fitting ears, fast-moving pupils that seem to jump out of his eyes, and canine teeth that are long, deformed, and protruding. He will have a humped neck, thin legs, bent feet with high arches, and six fingers on each hand. He is related to the fox, wolf, snakes, and tigers.[24] Deceitful persons display animated movements to the hands while they talk and are related to foxes, Ethiopians, women, and serpents.[25]

In general, we may observe that beauty, justice, and harmony are allied with a well-proportioned physique, while ugliness, evil, and discord are associated with bestiality, femininity, and duplicity, as evidenced in monstrously deformed bodies. The good news, however, is that regulation of the humors through diet, change of locale, and medications based on ingestion of the appropriate animal parts can bring about changes in temperament.

Book 4, for example, gives instructions as to how "Savage Men, leaving behind their feral habits, may become docile and pleasing."[26] Savageness is born of living in dry, harsh mountain locations that produce dry, bitter, and wild foods such as those where nomads, Sciites, and similarly savage peoples live. These rude primitives can be brought into civilization by large quantities of humid foods, since "small amounts of nourishment dry up the body and make savage [insalvatichisce] one's customs." Too much venereal activity also dries up the body; hence they should abstain from the exercise and movements of sex. They should avoid staying awake for long periods of time, fears, and similar emotions, indulging instead in otio or leisure, abundant sleep, food, and drink. Thus ends Della Fisonomia di tutto il corpo humano.

THE BEAST WITHIN: THE PIG, THE COW, THE CROW, THE MONKEY, THE PYGMY

Figures 9 to 14 are taken from three different editions of Porta's Physiognomy, ranging from 1616 to 1668.[27] These illustrations show better than any words how literally the traits of animals were believed to emerge from the human body. The first shows how toes that grow closely together resemble the cloven hoof of a pig. While Albert the Great interprets tight toes as demonstrating an envious character, Porta disagrees: having observed in a number of friends a membrane growing between the toes all the way up to the first knuckle, he says, "I knew them to be

FIGURE 9
The pig's tightly joined toes compared to the foot of a man.
From Porta's *Della Fisonomia dell'Huomo,* 1652 ed., p. 318.
(Courtesy of the Bancroft Library)

extremely shy."[28] Nevertheless, such people will be ugly and lusty like pigs, with porcine, dirty habits and an ambiguous character.

Figure 10, from the 1616 edition, shows the plight of men whose knees collapse inwardly with outwardly turning feet. Aristotle, Africanus, and Gellius make no bones about the weak character of these fellows: they are simply effeminate, like a certain P. Gallo, "an effeminate man who walked with his legs and feet turned about. . . . [T]hey are related to women." Porta interprets these men a little more generously than his ancient sources: "I would say that they are tranquil and peaceful and modest, and I would compare them to the Cow, who walks in this manner and whose legs are turned in like this, and in common speech we say Bovine legs."[29]

Figure 11 shows how the curvature of fingernails refers to the crow, "by nature inclined to theft, because servants steal money and whatever things they find in the house and they put them in gardens or hidden places or in holes and they hide them."[30] If the nails are also narrow and long as well as being curved, an eagle or buzzard is likely to be present, indicating not only impudence but "an insensate, bestial man."

Porta's explanation is based on Galenic physiology, more specifically on the association between natural growth and quantity of heat: "Everyone

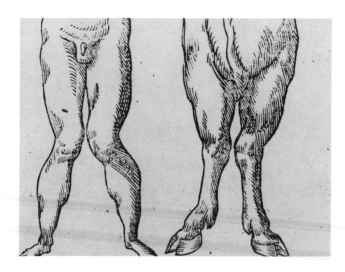

FIGURE 10
The collapsing legs of an effeminate man compared to bovine legs.
From Porta's *Della Fisonomia dell'Huomo*, 1616 ed., p. 164.
(Courtesy of the Bancroft Library)

knows that [since] the nature of nails arises from superabundance, and nature's progresses arise from heat, the narrowness of nails is followed by ignorance and crudeness, because in them gentleness is very feeble, nor can it make much progress by being able to dilate and become superabundant; hence those that have such little heat are ignorant and foolish because any coldness brings with it dullness and a crude wit and therefore those who have narrow nails will be ignorant and dull witted." And as irrefutable evidence of this theory, Porta goes on to cite "Plautus, speaking about a thieving cook, [who said]: *Go ahead and try to find a cook who doesn't have nails of an Eagle or a Buzzard.*"[31]

Figure 12 from the 1668 edition compares overlapped, scalelike nails with monkey paws. Elsewhere Porta explains that monkey characteristics are also evident from agitated body movements, especially bowing and scraping motions typical of dogs and sycophants, who "lower themselves and break their torsos in this way" in adulation and flattery. "I would compare them to Monkeys and would say that they are buffoons and malicious, those that always move about and twist their bodies and imitate human actions."[32]

Figures 13 and 14 come from different editions and are designed to illustrate different body parts. Figure 13 shows the physical characteristics of a lustful man, while Figure 14 illustrates the sidelong, slanting eyes of Venus, the personification of womanly lust, as she is being fondled by a satyr. An unknown owner of the book has traced in brown ink the missing pudenda of Venus, turning the engraving into an erotic invitation rather than a warning.

What strikes my eye is the similarity between the profile and body of Lustful Man in Figure 13 ("slender, nervous legs that speak of the generation of sparrows [or 'cunts' *passera*], hairy legs, fatty stomach and eyes, having a beard that curves up toward the nose and the circumference of the part between the nose and beard being concave. . . . the veins of the arms apparent . . . concave shining, not tearful, eyes that appear plunged into pleasure, while he incessantly moves his lips."[33]) and that of the satyr in Figure 14. Lustful Man seems caught in the act of stealing away from the scene pictured in Figure 14, one hand pointing down, the other splayed as if he had just fondled a woman's private parts. The spectator could easily imagine those hairy legs metamorphosing into cloven-hoofed donkey legs. The barrier between humanity and bestiality could not be more precarious than in these figures that illustrate the effects on the body of immoderate passions.

It would seem from our modern perspective that the presence of residual skin membranes between toes that point to cloven hooves, latent eagle nails, monkey paws, and satyric profiles would indicate some kind of evolutionary relationship between humans and animals. Our evolutionist paradigm teaches that human hands and toes evolved *out of* hooves and

FIGURE 11
The narrow, curving nails of a man compared to a crow's claws.
From Porta's *Della Fisonomia dell'Huomo*, 1668 ed., p. 174.
(Courtesy of the Bancroft Library)

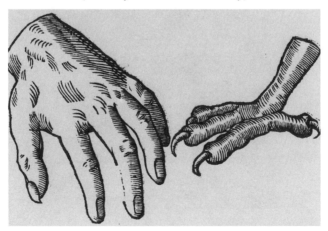

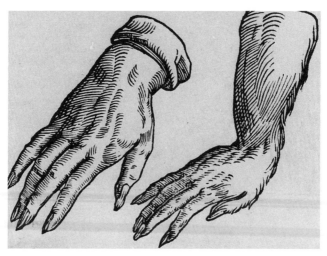

12

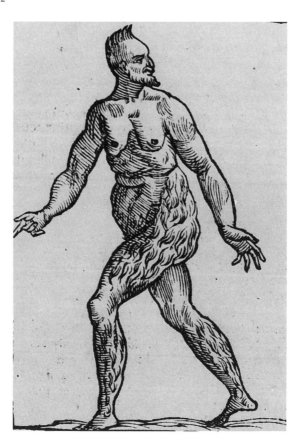

13

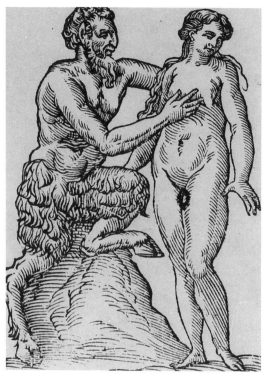

14

FIGURE 12
A monkey's paw compared to the imbricated, hard nails of a man.
From Porta's *Della Fisonomia dell'Huomo*, 1668 ed., p. 175.
(Courtesy of the Bancroft Library)

FIGURE 13
The Lustful Man. From Porta's *Della Fisonomia
dell'Huomo*, 1616 ed., p. 194. (Courtesy of the Bancroft Library)

FIGURE 14
Venus being fondled by a satyr.
From Porta's *Della Fisonomia dell'Huomo*, 1668 ed., p. 206.
(Courtesy of the Bancroft Library)

claws, just as we believe that dinosaurs might have evolved out of birds, and humans out of early primates. The vestigial similarities between human and animal bodies are evidence, for us, of this hypothesis.

But this interpretation would be anachronistic when describing the beliefs of physiognomy. Placing the body part of a man beside that of a similar animal is not meant to indicate a relationship in natural history, in a narrative of time. The links to be made in the discourse of physiognomy are, to use Roman Jakobson's terms, paradigmatic, not syntagmatic; they are metonymic, not syntactical. Man is never thought to develop *out of* animals over the course of time (a heretical notion, in any case, just as it is today for creationists); if this were so, the evolutionary theory would have to be reversible; that is to say, humans would have to be able to devolve back into animals. In a sense this is true, but the possibility for degenerating into the habits of lower beasts is no more likely than that of ascending through the intellect into the higher habits of angels.

The possibility for developing in either direction on the chain of being is the most convincing argument for free will. It is the battle between the rational soul and the animal soul that determines and maintains human form and character. In a sense, for a Christian, this is really the only battle that counts: since reason is the divine part of man, the spark of godhead situated in the head or heart is responsible for taming and domesticating, controlling and subduing the animal instincts of the body. The presence of "monkeyness" in a human, or "humanness" in an animal or plant points to a fluidity of being, a fluidity that arises from the interactions between material elements and spiritual elements.

A work published at the turn of the seventeenth century by the Neapolitan scientist Francesco Imperato gives a clear example of this notion. *Discorsi intorno a diverse cose naturali* serves as an addendum to Imperato's father's monumental *Dell'historia naturale* (Napoli, 1599), elaborating on controversial topics that the son feels require further explanation. Imperato wrote the book in order to respond to comments or questions elicited by visitors to his museum, some of whom questioned the authenticity of a number of objects in the collection.

The second discourse treats of "things produced by nature that have similarities and demonstrate the shape and appearance of many natural things."[34] Imperato begins by praising the usefulness of knowledge about natural things when applied for the benefit of human bodies. Contemplation of the Great Artificer's power in generating such beauty is aided by admiring the conformities or correspondences between land, air, and water animals, zoophyte plants, and rocks and minerals "generated by nature, in similarity to each other."[35] His overview includes similarities between plants that exhibit animal shapes; shells that mirror human hair; monkeys that imitate human forms, appearance, and gestures; and a stone naturally sculpted in bas-reliefs representing human hands and feet.

Often these jokes of nature lapse into deformity, giving rise to monstrous fruits, plants or animals with too many or not enough parts. Although Imperato alludes to these latter "brutes" as the monsters that "are reputed to be portentous," he does not seem to regard them with any sacred terror. On the contrary, he refers his readers to the "many examples of which can be derived from the accounts in the histories," thus placing his own discourse squarely between the scientific treatise of the Aristotelian sort and the wonders-of-nature genre following from Pliny.

The continuum of beings that playfully mimic each other's forms allows for an equally free play of monstrosity across the taxonomic categories that structured Imperato's theater of nature. By placing divinatory monsters in the same rubric as his description of, for example, the mandrake root (also called circean, perhaps, says Imperato, because the name is derived from the enchantress Circe, infamous maker of beasts out of men), Imperato would seem to imply that the forces of similitude and differentiation that nature has at its disposal have equal force and equal significance in the vegetable, animal, and human realms. The reference to Circe, to the world of literature and myth, adds language as another system of signs that also participates in the witty game of correspondences. Words and things, names and beings, all are able to refer to each other in meaningful patterns across the ontological orders. Monsters may be reputed to be portentous, but here, we sense, their true function is to provide a scientific concept of deviation from the norm, a morphological threshold of normality that provides a museological principle with which to justify the collector's inventory. *Monstrous* for Imperato is nothing more than a synonym for *unusual, exciting, thought-provoking.*

The author's explanation for monstrous deviation of forms in nature follows verbatim the precepts of Aristotelian embryology: either the maternal matter was lacking or excessive; the paternal seminal power was feeble; the place of generation lacked the proper qualities; the celestial influences were wrong; the imagination of the mother was uncontrolled. In any case, these extraordinary productions are evidence of God the Creator's power "to whom no thing is impossible no matter how very extravagantly we may consider it to be made."[36]

All this fits in well with the micro-macrocosmic model. However, the authenticity of his pygmy specimen, cause of much skepticism, is harder to prove. Imperato is forced to consider an evolutionary explanation based on biblical history, yet this historical narrative is continuously undercut by the co-presence of peoples of varying heights and statures, including contemporary dwarves and midgets. The fact is, peoples may get smaller, then bigger, then smaller again. Evolution has less to do with this question than the aesthetic biases of nature:

> The Lord God created our first Fathers from whom human generation originated; & in the progress of time the world abounded with people who were, & are of ordinary stature, & height; & even in the time of Noah, there were the Giants whose height was very out of proportion from the ordinary; why should nature not strive to generate Pygmies? who, being roughly three hands tall, are not so distant from the height of ordinary people; nor can they be considered very disproportionate in comparison with the height of Giants; besides, we see that Nature is more pleased by small statures than large ones; which is confirmed by observing the diminishing stature and height of humans, & little by little men were reduced to the lowest stature, which was in the first centuries, something that Homer was always complaining about; and along with height they lost years and health; and if Giants existed in many diverse parts of the world, we may surely say the same of Pygmies.[37]

Biblical and Homeric history show that we could just as easily become smaller or bigger as the climatic and social conditions may dictate, or as God wills. Variations in human size are more than anything proof of the creative powers of our Creator, who may amuse Himself as He sees fit, even by making odd-statured peoples. Imperato cinches his argument with the example of contemporary Japanese people, who, like the Pygmies, live in continuous isolation in harsh climates without commerce with other, taller populations. The explanation for the smallness of these peoples is similar to the explanation presented in Porta for wild men: the food produced in their sterile environments lacks the proper qualities to produce abundance. In any case, given the multiple factors at work, there is little that can be predicted regarding the future of human evolution.

Imperato finishes the fifth discourse with an eyewitness account of a short, primitive people glimpsed by an Augustinian monk from Lapland whose ship was blown off course:

> A few years ago a very Reverend Augustinian Monk from Lapland told me that on a voyage by way of Flanders, a sea storm brought him to a place near the Island of Islan . . . ; but the violence of the wind transported him further, where he found a safe refuge, and there he stayed for a few days, and wanting to identify this place, if it were inhabited, and what it was called, and if it were an Island or a

continent, he found it deserted, & uninhabited; so, he pushed on further inward, encouraged mainly by having found some small cabins, made of wood and others of fish bones from the Cretaceous species, interwoven with fine craftmanship with certain unusually large bulrushes; but while walking around he found a bush, inside of which he saw two people, who at first glance appeared to be shepherds, clothed in extremely white skins; and because they quickly fled, he was not permitted to carefully observe their clothing, nor their appearance and height; but their stature seemed to him to be very small.[38]

We may note from this account that while the shepherd couple is of short stature, they are understood to be culturally evolved: their clothing and housing are said to be made "con bello artificio." There is no hint of monstrosity or bestiality attached to their unusual physique.

This observation would indicate that the barrier between human and animal is not based simply on physical evidence; if it were, the pygmy and the giant would have to be excluded from the ranks of humanity. Rather, it is a combination of physical and spiritual qualities and the interaction between the two that both define the human being and constitute the principles of physiognomy. In the play of differences and similarities displayed throughout the continuum of natural beings, the physiognomist, like the natural scientist, seeks to uncover the range of human experience glimpsed in the lineaments of animality. And we shall see that while a pygmy is an authentic member of the human family, a woman is not.

"WOMAN IS A MONSTER OF NATURE"

Having familiarized oneself with the general principles of physiognomy, how would one go about applying them to specific individuals who present complex combinations of attributes? For this the practitioner must turn to the 1670 edition of Cornelio Ghirardelli's *Cefalogia Fisonomica*, a marvelous practical manual comprising 624 pages, divided into ten books with ten chapters in each, illustrating the principles of physiognomy. Each body part, following pseudo-Aristotle's lead, comprises one book: hair, forehead, eyebrows, eyes, nose, mouth, chin, ears, face, and head. What makes Ghirardelli's manual so useful is that he includes a picture of the particular kind of body part he is discussing, so that we are regaled with "one hundred engravings of human heads which can be seen in this Work, from which are demonstrated various tendencies of Men and women by means of many signs and conjectures." In addition Ghirardelli has collected "just as many sonnets by diverse and excellent Poets and

FIGURE 15

A portrait emblem for the woman with a short, upturned nose.
From Ghirardelli's *Cefalogia fisonomica*, p. 334.
(Courtesy of the Bancroft Library)

Academicians." These augment the Latin epigrams that adorn each engraving. Ghirardelli's texts are followed by "Additions to each Discourse
by Inquieto, a member of the Vespertine Academy."[39]

The result is a full-spectrum look, combining rhetorical and scientific
discourses, at this method of reading the signs of the human body in
terms of animal and celestial characteristics. As we would expect, monstrosity arises whenever the perfect, male body deviates either into femininity or deceitfulness, figured in bestial traits that poke through the
mask of humanity. We will limit our exposition to two such examples:
women with small, upturned noses and those with small heads and
slender necks.

Figure 15 shows how "a woman with a small upturning nose is deceitful, arrogant, and spews forth curses." The sonnet by Sig. Tobia Tobioli
elaborates on this theme, basing its simile on a monstrous figure, the
siren:

Men, o you who never want your path to be deviated [lit. your foot to be twisted] by women, immersed, nay submerged all the time in that Aegean (Sea), that on the outside [is] alluring, impious Siren, and inside [is] ill-feeling spleen. Flee, incautious ones; there is the seat of Love; not the kind of Love that resides in Heaven; but truly the kind that always stabs through the heart. [They are] full of deception, empty of faith. Flee, even more than other women, alas, flee those whose nose is upturned and short; for they are vain, lascivious and far too impudent. They amuse themselves in boasting, they are bound to deceits and fury, they speak evil of others, but always wrongfully.[40]

While we might find a small, upturned nose to be scant evidence of character in a woman, the siren topos alerts us that women in general are not to be trusted. As the anonymous "Inquieto" explains in the tortured prose of his additional commentary, all women are the same when it comes to their affects, inasmuch as they all lack moderation: a woman either loves disproportionately or hates mortally; she knows no median, and is born to deceit and fraudulence.[41] The reason for this is a well-known fact: "Woman is a monster of nature; she is an imperfect man, as many Learned writers are pleased to determine, which we may deduce from all her parts."[42]

Monstrosity is thus attributed both to imperfection, in the sense of deviation from the perfect norm, and to disproportion. It is located both in the passions, which are excessive, and the body parts, which do not exhibit the proper male ratio. Monstrosity, like ugliness, is a lack of proper means and proportions between parts, the principles that define classical notions of beauty. Hence, while a woman's physique might superficially evidence characteristics of beauty, internally she is flawed and deformed by virtue of her gender: her passions are out of order, she loves and hates with no measure, she is an enemy to moderation. This duplicity or contradiction between inside and outside is also characteristic of the siren, an alluring figure who universally signifies the dangers of the flesh and of pleasure in general. Monstrosity is a complex combination of physical characteristics and spiritual flaws attesting to the close relationship between body and soul, matter and spirit.

The last section of Ghirardelli's compendium discusses the case of women with small heads and slender necks. The explanation for her ensuing feebleness, malnutrition, and badly formed body is quite explicit, providing a preview of the physiology of animal spirits. Just as a vessel of liquor determines the quantity of the spirits that it contains, the volume of the brain depends on the dimensions of its container. Since women's heads are small, the material contained within them is proportionately restricted. The narrowness of a woman's head suffocates and almost extinguishes the animal spirits "and from this is born feebleness." Now, because of these enfeebled spirits, unable to properly carry out their functions, the soul becomes hazy and dubious, "unable to discern the truth of things." No wonder then, concludes Ghirardelli, that women are typically afflicted by a vacillating will. No sooner does their soul "take hold of any resolution but that it wonders if it should have done the opposite (a passion typical of Women)."[43]

The discourse that follows is a curious one drawn from the personal experience of the author. He begins by listing the characteristics of "similar Women we have observed, both in the shape of their body as in the qualities of their spirit." Such women are tall with slender, nervous limbs. Their eyes are tan colored, their brows smooth, their noses—as we would expect—slight and upturned, with a very large mouth and quite a big lower lip. Their ears are also big and, for the most part, these women have an olive complexion. This much falls in line with the precepts of physiognomy; what follows, however, seems quite out of keeping with the scientific discourse. The author takes on an almost querulous tone as he describes the perils of these seemingly emancipated, and most certainly literate, creatures: "They are curious for profane books, especially ones about love, and this derives from the libidinous appetite which prevails over them. They are desirous of novelties; they are constantly wanting to change their clothes; they are inclined to leisure, delights, and pleasures, to music, sounds, and games, not to womanly pursuits and domestic cares; they are generous in giving. . . . They enjoy being served and making frequent use of courtesies."[44]

All this would be tolerable were it not for the fact that this kind of a woman is so eloquent that she runs circles around her male admirer, who in the following scene breathlessly records his enchantress's witty sayings in a notebook. The seductive song of these small-brained women signals the birth of yet another siren, which brings the last page of the *Cefalogia Fisonomica* to a close:

> Because they are very beautiful speakers, in such a manner that they would hold at bay any more retiring and devoted spirit & would soften only with words the most hardened heart, saying at times such graceful and stimulating conceits that a friend of ours would tell us that he often had dealings with one of these women simply because he was infatuated by her delightful manner of speech. In fact, sometimes he would take note of some of her clever remarks in order to make use of them for his own literary purposes. But the truth is

that he himself swore that he had bought that Woman's words with his own blood, because allured by them, like a little Bird caught by the simulated whistle of a Bird-catcher, he fell into the net of love, and the fall was such that he was unable to extricate himself without remaining maimed in his life, his property, and his reputation. Let this be an example for all men to live as far away as possible from these false Sirens.[45]

Being small brained and, thus, confused by nature does not seem to stop some women from displaying their brilliant intellectual wit. Inexplicably, the author of this story does not seem to pick up on the obvious contradiction between limited female physiology and these particular women's superior eloquence. He does take for granted, though, that plagiarizing from women does not count as such.

Apart from having such an inherently contradictory nature, the problem with these women is that the brighter they shine in company, the more powerful, dangerous, and monstrous they turn out to be. Circelike, they make *augellini,* or little birds, out of their admirers, who fall haplessly into their nets of delightful speech. These clever girls manage to be, at once, both bird-brained and bird catchers.

Given that the sirens of Homeric yore were traditionally depicted as half birds (a woman's head on a bird's body that enchanted its victims with its song), this metaphor of transformation is striking for its role reversal. In this anecdote, the male victim is the one that becomes a little bird. It is as if the male admirer is transformed into what he most fears, and yet desires, to be: powerful, eloquent, sirenlike. His literary success depends on it. These alluring small-headed women with slender necks cast spoken nets around their twittering prey, making them pay in blood for their surreptitious thievery. But the truth is, as frightened, maimed, and abused as this particular literary vulture is by his enchantress, he wants nothing more in the end than to be like her.

CURING THE SOUL

While one might not be able to cure oneself from being female, most every other type of deformation could be tempered. The sophisticated diagnostic and prognostic apparatus of physiognomy was developed not only to protect oneself from deceitful courtiers; it was primarily conceived as a medical art, as a self-help science that treated the passions along with the body. By carefully observing the eyes and the face, a student of physiognomy can diagnose particular vices by their physical signs. By purging or medicating "those defective humors" we may liberate ourselves from that vice. "Therefore," concludes Porta, "let us all most willingly seek out this science, & embrace it with all our hearts, as something that is truly our own, and that treats of us ourselves."[46]

As the repeated play of double reflexive pronouns reveals, the more you look at yourself without, the better you know yourself within: "acciò che noi stessi di noi medesimi diventiamo Fisonomi [so that we, ourselves, become Physiognomists of us, ourselves]." The mirror, principal device for "knowing thyself," is often recommended as an instrument for physiognomists intent on self-improvement. As the practitioner stares at his image in the surface of the glass, within and without merge, body and soul become images of each other. Like Oscar Wilde's portrait of Dorian Gray, the body changes and expresses itself according to our manners and behavior. Both Socrates and Seneca believed that "man may mirror himself," explains Porta, and "the Philosopher Socrates used the mirror for the good teaching of habits": "When someone looks in a mirror, and sees himself well formed by nature, he may see to it for the future that he not stain the beauty of his body with the ugliness of habits; in the same way, seeing the body ugly, he may see to it with all his effort and diligence that by means of virtues, he may medicate and heal the ugly signs of the body."[47]

The beast within is perfectly imaged in the contours of the flesh, just as deformity of soul, the ugliness of vice, can transmogrify a beauteous semblance into a hideous monster. To put it simply: "With illness of the body only the soul changes, since it is unable to perform its task, and the man who suffers the illness is not the same, but becomes other."[48] By "other" Porta does not mean simply that the patient feels out of sorts; the patient actually becomes other than what he is, a man: "Due to great and extraordinary passion many men have become worse than beasts. . . . From the illnesses of Cynanthropy and Lycanthropy, many change into dogs, their eyes become fiery, with threatening teeth and a sharp nose; they go out at night, prowl around graves, nor does one hear anything from them but howling, snarling, and other things."[49]

Granted, lycanthropy (werewolf disease) is an extreme example. Nevertheless, the principle holds that when the pituitary organ is affected, choleric and bitter humors roam about the body. The rising vapors penetrate the innermost places of the soul. The vapors force the soul to give up on being what it was and to become instead bold, shy, crude, or deceitful. That is why, when it comes to illnesses of the soul, "they are healed by medicating the body."[50]

We recognize in this line of thought the Aristotelian assumption that the arrangement (the disposition, ordering, or placement) of matter is what gives rise to a particular kind of being (spirit, soul, consciousness). In his *Fisionomia Naturale*, Giovanni Ingegneri, Bishop of Capo d'Istria, explains the relationship using a sartorial metaphor that we have encountered before: "The soul, which gives being to the body, & lives inside it, is herself the one who arranges it, not really like her house & dwelling, but she adapts it to her fit, just as a tailor adjusts a skirt to another person, or a stocking." "And an animal would be imperfect if there were not found a true correspondence between soul and body; because matter would not be obeying form."[51] Harmony between soul and body is the exact relation between form and matter, male and female, master and servant, shuttlecock and weaver. Monstrosity or imperfection results from the unruliness of matter when it disobeys its ruling form.

The notion of instrumental use, or final cause, completes this explanation. Porta explicitly refers to Aristotle when he explains that the final cause of a thing, its final purpose, is what determines its material arrangement. "The action of sawing was not made for the saw, but the saw in order to saw, hence the body was made for its offices." Therefore "nature provides a body proportioned according to the actions of the soul" and "the whole body was created by nature for some most excellent action."[52] It follows that "the arrangement of the body responds to the powers and virtues of the soul; moreover, the soul and the body love each other with so much correspondence, that one is the cause of joy and pain in the other."[53] "From which I believe," says Porta, "that whoever is missing any principal member, is monstrous, and ill-fortuned, & that he also lacks something of foresight, & prudence; & those of evil habits, & little prudence are always subject to harmful things."[54] Thus, deformation, evil living, and bad luck are intimately associated with each other and they all indicate a lack of balance between the animal part of man, his body, and the angelic part, his intellect or soul. "Know," Porta intones, "that man was located by God in the confines of the supreme intelligences and the beasts; because through our intellect we resemble the one, and through our senses we resemble the other."[55] When a man falls to evil vices, such as eating human flesh, having intercourse with his daughters, or killing his sons or his father, says Porta, then he is actually worse than a beast, because an animal does not have the privilege of exercising his free will, of using his reasonable faculties in order to shun vice.

Accordingly, a man with the characteristics of a malicious beast resembles a bear: thick hair, redder than normal, hairy cheeks, hairy torso, hunched-up shoulders, small, corpulent feet, long, twisted, narrow nails with short, fat fingers. The verses that accompany this description testify that such a man is unsuited for the proper actions of humanity; he is a useless burden: "Che altro è l'Orso che un disutil peso. / Ferocità di una ben pazza mente. [What else is a Bear than a useless burden. / Ferocity of a truly crazy mind.]"[56]

While vice can drag us down a notch or two from our human status into the animal kingdom, virtue is an equally potent force that can uplift us a step or more up the chain of being. The hero, directly opposed to the malicious beast described above, inhabits a body similar to the angels because he fulfills a useful function to mankind. Those filled with heroic and divine virtue are semi-gods, Porta says, not because they look like angelic intelligences, but because they advance the human condition through their moral superiority.[57] By examining oneself in a mirror for signs of arrogance, envy, and other moral illnesses, and by adequately treating our infirmities according to the doctrines of physiognomy, we are offered the possibility of choosing who we want to become: foolish, despicable beasts or heroic semi-gods.

Man the Machine

It seems appropriate to begin this section with a reminder regarding the difficulty of historicizing one's choice of texts and themes when the subject matter is still very close to heart. The science of physiognomy provides a well-defined discipline from which it is easy to identify the old epistemological foundations passed down from classical times. The new medicine that appeared in the seventeenth century is more difficult to define: it is not contained in any one school of thought, nor is the concept of the body as machinelike entirely accepted by any one writer. Furthermore, we are still presently living and discovering the practical and philosophical implications of imagining ourselves as nothing more than complex machines mysteriously endowed with consciousness.

The question of whether matter is purely passive or whether it is endowed with its own energy, and the problem of whether animal behavior is determined purely by external stimuli or whether it depends on perceptual and deliberative capacities similar to human ones, are in the process of being reconceptualized even today.[58] This is partly due to new discoveries regarding the chemical bases of neuroreceptors and neurotransmitters: according to the latest medical research, subjective consciousness and objective perception are intimately fused throughout our whole physique in the form of chemical and hormonal components. Any distinction between body and mind, between fact and perception, becomes more nebulous and perhaps even outmoded to our postmodern understanding. Now, as in early modern Italy, *body*, *mind*, and *soul* are terms that are vigorously debated both in academic and popular culture.

In some ways we are living the end of the paradigm that began with the application of mechanical laws of nature to human physiology. As our conception of ourselves merges with the information and biotechnologies that we have created (necessarily in our own image), the metaphors we use reflect who we think we are. Networking, virtual reality, cyberspace, genetic coding—all suggest possibilities for redefining the physical and spiritual potential of humanity that leave mechanistic models behind.

Let us remember, then, that Western culture has suffered from what Joseph Needham has called "a schizophrenic character or split personality": the tendency to conceive the totality of nature either in terms of vitalistic, theological idealism or in terms of mechanistic materialism. The writers we are about to discuss sought to overcome that dichotomy. Theirs were local conflicts that were symptomatic of a much broader war of ideas, one that was decisive in establishing "matter," ruled by mathematically expressible laws, rather than moral virtue, as the most popular and descriptive tool for advancing the human condition.

Between the seventeenth and eighteenth centuries a gradual and complex shift in paradigms took place in European medicine: man the microcosm gradually transformed into man the machine. The narrative I am about to tell regarding that shift centers on one aspect of mechanism alone: the transformation of animal spirits contained in the blood and nerves into purely corporeal components of a fleshy automaton. The investigation into the nature of animal and vital spirits along with their role in creating the passions of the soul was the precise area of research in which questions concerning the nexus between matter and spirit were debated.

Since I believe that monstrosity is defined by natural or unnatural relations between inanimate matter and an animate life force, the choice of focus is clearly dictated by my theme. If I am correct in seeing inanimate matter moving of its own accord as a fundamental violation of natural law and, thus, defining the essence of monstrosity, then the new mechanical medicine was precisely the locus where man the machine became irrevocably monstrous in his own eyes. Due to length considerations I will limit my discussion in this section to William Harvey, René Descartes, and Giovanni Borelli. They are, I believe, the most important authors to pioneer the new mechanistic paradigm for the human body. Not only were these scientists the most important contributors to the European process of automatizing the body, they were also the main sources of influence and information (via the Academy of Investigators) regarding human physiology for Giambattista Vico, our focus in the next chapter.[59]

THE HEART OF CREATURES

The dedication of William Harvey's *Anatomical Exercises . . . Concerning the Motion of the Heart and Blood,* first published in Latin in 1628, reads as follows:

Most Gracious King,

The Heart of creatures is the foundation of life, the Prince of all, the Sun of their Microcosm, on which all vegetation does depend, from whence all vigor & strength does flow. Likewise the King is the foundation of his Kingdoms, and the sun of his Microcosm, the Heart of his Common-wealth, from whence all power and mercy proceeds. . . . You may at least, best of Kings, being plac'd in the top of human things, at the same time contemplate the Principle of Mans Body, and the Image of your Kingly power.[60]

Harvey (1578–1657) had studied anatomy in Padua with Fabrizio d'Aqua-pendente, who had produced marvelously accurate drawings of the valves in the veins, failing, however, due to his loyalty to Galen, to move beyond descriptive anatomy. Nine years of experimentation, ocular demonstration, and frequent vivisection leading up to Harvey's discovery did not dissuade the English physician that nature was a purposive entity: "So Nature being perfect and divine, and making nothing in vain, neither gave a *heart* to any where there was no need, nor made it before there was any use for it."[61]

The importance of the heart to the organism, one feels in reading his little treatise, is as much cosmological as it is physical: "The *heart* is as it were a Prince in the Commonwealth, in whose person is the first and highest government every where; from which, as from the original and foundation, all power in the *animal* is deriv'd, and doth depend."[62] The heart is to the body, as the prince is to the state, as the sun is to the universe. In a well-ordered state, public and private spheres are well delineated. Just so, the heart also acts as a treasury, distributing and dispensing nourishment to the members. Its impulse is a violent one.[63] This correspondence between hierarchy of organs and hierarchy of state recalls the Platonic precepts of justice with which we opened this chapter.

Despite his discovery of the circulation of the blood, Harvey was not a mechanistic thinker.[64] His use of numbers, measuring the output of the heart and counting its beats per minute, indicate a quantitative aspect to his mode of investigation, and though he does compare the two motions of the ear and the ventricle of the heart to the working of an engine, the simile strikes one more as a convenience of expression, as an apt comparison, rather than a paradigmatic model for investigation.[65]

Harvey is credited with founding modern physiology and biology, and his importance to comparative anatomy, studying the movement of the heart and blood in lesser animals and man as an animal, is undeniable. Yet, as Walter Pagel shows in his article "The Reaction to Aristotle in

Seventeenth-Century Biological Thought," Harvey's self-confessed debt to "the divine Galen" and to the main tenets of Aristotelian physiology are patent. The preeminence of the heart as the origin of heat, and thus of the life principle, is Aristotelian: "There must needs be a place and beginning of heat, (as it were a Fire, and dwelling house) by which the nursery of Nature . . . from whence heat and life may flow. . . . And that this place is the *heart*, from whence is the beginning of life, I would have no body to doubt."[66]

Indeed, the very notion of circular motion Harvey derives from Greek theories of *circumpulsion*. In this passage, Harvey describes how he came upon the model of a closed, ecological circuit for the workings of the body:

> I began to bethink my self if it might not have a *circular motion*, which afterwards I found true. . . . Which motion we may call *circular*, after the same manner that *Aristotle* sayes that the rain and the air doe imitate the motion of the superiour bodies. For the earth being wet, evaporates by the heat of the *sun*, and the vapours being rais'd aloft are condens'd and descend in showrs, and wet the ground, and by this means here are generated, likewise, tempests, and the beginnings of meteors, from the circular motion of the *sun*, and his approach and removall. So in all likelihood it comes to passe in the body, that all the parts are nourished, cherished, and quickned with blood, which is warm, perfect, vaporous, full of spirit.[67]

As we read in the above passage, Harvey characterizes the blood as "warm, perfect, vaporous, full of spirit." He further characterizes the blood as "being fraught with spirits, as with balsam." But it is the *motion* of the heart that imparts warmth and, hence, vitality to the blood: "For we doe see, that by *motion*, *heat* and *spirit* is ingender'd, and preserv'd in all things, and by want of it vanishes."[68] That is to say, without the violent impulse of the heart, blood would be nothing but matter without motion, a lifeless thing.

In a later book, *On Generation* (1651), however, Harvey changes his mind and disagrees with Aristotle by placing the heart, the engine that imparts motion, second in dignity to the blood. He does so referring to the authority of the Pentateuch: blood is the residence of life "because in it life and the soul first show themselves and at last become extinct."[69] What is pertinent to our discussion is not the preeminence of one or the other but rather the exact nature of the blood: as the substance that mediates between matter and spirit, its character is crucial in defining the source of life and hence the character of the human body itself. Pagel is cautious in interpreting this later identification between blood and soul, preferring to understand "soul" as "nothing but the natural function inherent in blood which acts as the material substratum necessary for the appropriate effects to be obtained in physical life." He concludes that Harvey's views "were anti-materialistic in the true Aristotelian tradition." Equally Aristotelian is Harvey's belief in "the immanence of life and function in organized matter, neither of which can exist independently of each other"[70] and which, as we have seen, is the basis of the old science of physiognomy.

The birth of mechanical models for the human body and the ensuing split between mind and matter, body and soul is not to be found in the man who discovered the circulation of the blood. Likewise, there is no hint of monstrous deformation in the malfunctioning of the organism. Harvey sees disease rather as being due to "cacochymie, or abundance of crudities." The link between mind and body is a wholistic one.[71] Man the machine has been hinted at but is still a notion.

ANIMAL SPIRITS

In the fifth part of his *Discourse on Method,* René Descartes (1596–1657) provides a summary of the middle section of his projected work *Le Monde,* which was never completed. The entire tripartite work was to contain a description of the stars, the heavens, and the earth, as well as bodies on the earth. *L'Homme,* a treatise on the principal functions of man, was meant to be sandwiched between one on light and another on the soul.

In late November 1633, having heard of the burning of Galileo's *Dialogho sopra i due Massimi Sistemi,* Descartes writes to Mersenne that the events in Rome "have so strongly shocked me that I am almost resolved to burn all my papers, or at least to not allow them to be seen by anyone." He goes on to explain that if the hypothesis concerning the movement of the earth is no longer permitted to be taught publicly, and if he is thereby forced to declare its falsity, then "all the foundations of my philosophy are also [false]. . . . And it is so tied in with all the parts of my treatise that I would not know how to detach it from them without rendering the rest completely defective." He would rather suppress the whole work, he continues, than distort it to fit with the censors' expectations.[72] So it was that the treatise *On Man* appeared only in a truncated form in 1637 in the French version of the *Discourse on Method;* in 1662 it appeared posthumously in Latin, and in 1664, in French.

Much earlier, by November or December of 1632, Descartes had already dismissed any influence from Harvey's work. In a letter to Mersenne he writes:

I will speak about man in my *World* a little more than I had thought to, since I am undertaking to explain all his principal functions. I have already written about those that concern his life, like digestion of foods, the beating of the pulse, the distribution of nourishment, etc., & the five senses. I am now dissecting the heads of various animals, in order to explain what the imagination, memory etc. consist of. I saw the book *de motu cordis* that you had spoken to me about, & find my opinion little different than his, although I only saw it after having finished writing about this matter.[73]

It is not easy to pin down Descartes's beliefs on the nature of the soul. In the *Discourse on Method* Descartes makes clear that any physiology must begin from the premise that God created the human body at first without any vegetative or sensitive soul. A heat without light kindled in the heart replaces these entities, and only later is a unitary, rational soul placed in the body. Although "our soul is of a nature entirely independent of the body," "it must be more intimately joined and united with the body in order to also have sentiments and appetites like ours, and so to make up a real man."[74]

In the *Sixth Meditation* Descartes declares the human essence to be that of a body which thinks. The "I" of thought is entirely distinct from the body, and yet thinking has its origin in and depends upon the union and apparent fusion of the mind with the body. The idea of being able to separate the thinking "I" from the body is ludicrous. The *whole* mind seems to be united with the *whole* body, and not simply like a pilot who resides in a ship: the mixture is so blended that something like a single whole is produced. He concludes that the nature of man is faulty and deceptive precisely because of the fact that man is composed of mind and body.[75]

What, then, unites this awkward hybrid? For the answer to this question we must turn to his work on physiology proper. Let us imagine that we were to construct a man similar to the one that God constructed. *L'Homme* begins with this conceit, under the title "On the Machine of his Body": "These men will be composed like us, of a Soul & Body; And I must first describe to you separately the body, then afterwards the soul, also separately: And finally I will show you how these two Natures must be joined & united, in order to compose men who resemble us."[76] He continues by positing that the body is nothing but a statue or earth machine that God formed to our semblance in such a way that it would imitate our functions "that may be imagined to proceed from matter, & to depend solely on the arrangement of organs."

We see watches, artificial fountains, windmills, and other similar machines, he says, which, even though made by men, still have the power to move of themselves.[77] Since that which moves of its own accord is animate, then to define the principle of motion, whether in the heavens or in the body, is to define the location of soul. In a continuous tradition of inquiry, from Aristotle to Giovanni Borelli and beyond, the question of *de motu animalium*, animal motion, is crucial to defining the principle of animation and of life itself. As we have seen, it also serves to distinguish between brutes, human beings, and machines.

By thinking of the human body in terms of hydraulic machines, such as "the grottos and fountains that are in the gardens of our Kings," Descartes constructs his theory of movement based on nerve channels conceived as tubes and pipes. If you look at these automatons you will see "that the sole force by which water moves itself as it issues from its source is sufficient to move various machines, & even to make them play several musical instruments or pronounce a few words, according to the arrangement of pipes that channel [*conduisent*] the water" (see figure 7 in chapter 2). Not by chance, in describing the coursing of spirits through these hydraulic pipes, spectator-readers are regaled with a description of "a marine monster who will vomit water against their faces."[78]

In a section on joy, sadness, and other internal sentiments, Descartes draws an extended analogy between an organ and the human body. The coursing of air into the pipes is compared to the heart and arteries "that push the animal spirits into the concavities of the brain of our machine."[79] External sense objects move particular nerve endings, which in turn enter into one or another pore of the brain. "And just as the harmony of organs hardly depends on the arrangement of their pipes that we see from outside, nor from the shape of their wind portals or other parts, but only on three things, that is, the air that comes from the bellows, the pipes that render the sound, & the distribution of this air in the pipes; In the same way I would like to inform you that the functions . . . depend only on the spirits that come from the heart, the pores of the brain through which they pass, & the manner in which these spirits are distributed in these pores."[80] All the humors or natural inclinations of the soul may be explained by the quantity, coarseness, degree of agitation, and uniformity of the particles of animal spirits. For example, if these spirits are more abundant than usual, goodness, liberality, and love will ensue; if the particles are stronger and bigger, we will feel confidence or courage; if they are more agitated, desire and diligence are the result; uniformity of motion gives rise to tranquillity of the spirit.[81]

Explaining emotions and character through the simple variance of particles is a significant departure from the precepts of physiognomy and Galenic humors. Any change in blood causes a change in the animal spirits and subsequently to one's mood. The quality of blood depends on the nourishment passed through the stomach through the veins, which, mixing with the blood, communicates its qualities and makes it coarser.

Blood that has circulated a number of times through the heart becomes more subtle. Respiration through the lungs causes the air to mix "en quelque façon [in some way]" with the blood, revivifying and agitating the spirits.[82]

If blood and its spirits are the sole agents of animation in the human body, they must also be the joining and uniting principle between matter and soul. Let us examine how and where the animal spirits are produced:

> Now this blood contained in the veins has but one visible passage by which it can exit, to whit, the one that leads into the right chamber of the heart. And know that the flesh of the heart contains in its pores one of those fires without light . . . which renders [the blood] so hot and so fiery that as it enters into one of the two chambers or concavities that are in the flesh, it immediately puffs up and dilates there. . . . And the fire that is in the heart of the machine . . . serves only to dilate, heat up, & make subtle the blood, which continually falls, drop by drop, by way of a tube in the hollow vein, into the right side of the concavity, from whence it is exhaled into the lung; & from the vein of the lung . . . into its other concavity, where it is distributed throughout the body.[83]

How are we to imagine this "fire without light" that inhabits the fleshy pores of the heart, puffing up and dilating, heating, and nebulizing the blood? Not motion, then, but some mysterious, innate property of the heart is what imparts spirit to the blood. Yet, says Descartes later, this fire without light is no different from any other fire that inhabits inanimate bodies. It must be an ordinary fire in order for the "machine de son corps" to be what it is: an ordinary machine.

The treatise ends with a paragraph based on the same hypothetical conceit that began Descartes's inquiry into man:

> I desire that you should consider after this, that all the functions that I have attributed to this Machine . . . should imitate as perfectly as possible those of a real man; I desire, say I, that you should consider that these functions follow completely naturally in this Machine, from the sole arrangement of its organs; no more or less than do the movements of a watch, or another automaton, of that of its counterweights & its wheels; So that there shall be no occasion to conceive in this machine any other kind of vegetative nor sensitive Soul, nor any other principle of movement & life, than its blood & its Spirits, agitated by the heat of the fire that burns continually in its heart, & which is of no other Nature than all the fires that are in Inanimate Bodies.[84]

Clearly, in this posthumous work Descartes takes as his premise that the functions and organization of the body may be analyzed *as if* they were those of a machine, a watch, or another automaton. He does so in order to demonstrate his novel hypothesis: that there is no need to posit an immaterial force to explain any aspect of human physiology. Everything in nature, including man, may be investigated according to the same descriptive apparatus, and we may henceforth dispense with qualities, substantial forms, occult virtues, even the tripartite soul. Matter is sufficient to describe the material world, of which man is a part. The principle of movement and life is attributed solely to blood and animal spirits, which are posited as the material substratum of a unitary soul.

Now if such a machine can be imagined, then Descartes has proved his point with nothing but a sophisticated thought experiment. He has not actually stated that the human body is made of such stuff. He has simply demonstrated that it *could* be imagined as a machine. To assert anything more would have led to accusations of impiety, denial of the immortal soul, and so forth. The machine conceit thus serves as both a heuristic and an apologetic device.

We must conclude that any notion of dualism in Cartesian thought between body and mind, matter and spirit must be of a qualified, even ambiguous sort. A far more accurate description would be that of Aristotelian immanence: life and function arise from the arrangement of matter. Unfortunately, in the *Discourse on Method* Descartes denies this view: while the correct disposition of matter might allow us to create machines and automatons that are virtually indistinguishable from animals, a machine, no matter how complex, even if it could talk and write, could never appear to be human; there is an inviolable limit to artificial intelligence. "Although machines could do many things as well as, or perhaps even better than us, they would infallibly fail in certain others, by which we would discover that they did not act from awareness or knowledge, but only from the disposition of their organs."[85]

In spite of all the appeals to automatons and machines that pepper his writings, Descartes clearly states here that human beings are not simply complex machines. Machines lack a rational soul, comparable to the engineer who designs and runs the automatons in royal gardens. Even to characterize Descartes's physiology as "mechanistic" would be too simplistic.[86] I leave it to the reader to reconcile the above observations with the passages I have quoted. The only way to characterize Descartes's statements on dualism and mechanism without simply pointing to their contradictory nature is to assume either that fear of censorship and persecution led him to equivocate or that from year to year his views changed. If we compare his posthumously published physiological writings of the early 1630s with their synopsis in the *Discourse on Method* (1637), no reconciliation is possible.

DE MOTU ANIMALIUM

The mechanistic conceits of the French philosopher are not sufficient to explain the advent of the machine-body. For that we must turn to the Italian inheritors of Galileo, and specifically to Giovanni Borelli, in order to understand how the laws of nature became applied to the human body. Lorenzo Magalotti, secretary to the experimental academy of Florence, tells us that Borelli was apparently "an irksome man, and almost I said completely intolerable, but all in all he was a man of letters such as to make a court shine, because he had soundness and judgment."[87] Born in Naples in 1608, Borelli obtained the public lectureship in mathematics in Messina, Sicily around 1635. In 1658 he was given the chair of mathematics at Pisa. There he collaborated with Malpighi—who dissected live animals in Borelli's home for him—and became a principal force of the Accademia del Cimento, the experimental society under the patronage of Prince Leopold, brother of the Grand Duke of Tuscany, Ferdinand II. After a decade at Pisa and Florence he returned to Messina in 1667 and spent part of the summer at Naples repeating before the Academy of the Investigators the experiments on animal motion he had carried out in the Cimento.

Borelli was a student of Castelli—friend of Galileo's and instigator of modern hydraulics—and a large part of his work was a continuation of Galileo's research interests. Many of Galileo's lost writings relive in the manuscripts or published writings of Borelli, one of the founders of "iatrophysics" (the application of physics to medicine).[88] As Raffaello Caverni notes, "All of his experimental studies, even the most disparate in appearance, were united in a single intention, which was to apply Mechanics and Physics to the movement of animals." The publication rights for his magnum opus *De motu animalium*,[89] which appeared a year after his death, were vied for by the Royal Society of London and "several academies from the remotest parts of the world" as well as universities in Holland, France, and Italy who "asked the Author insistently for his work, to publish it at their expense."[90] The honor fell to Queen Christina of Sweden, friend and pupil of Borelli.

Borelli's aversion to the Cartesian approach to physiology was adamant and led to a break with Malpighi—honorary member of the Royal Society of London and founder of microscopical anatomy—who believed that the application of the new physics and mechanics should derive both from Galileo and Descartes. For Borelli, Descartes's insistence on wanting to accommodate facts to philosophical reasoning seemed nothing less than madness, fictions of the mind. Borelli specifically rejected Descartes's hypothesis, borrowed from Galen and maintained by all anatomists since then, that muscular motion derived from invisible expirations of the brain (the spirits composed partly of air that arrived to the muscles from the brain by way of tubular nerves). Borelli easily dismissed this fiction by observing that when a nerve is dissected there are no air bubbles in it.

He also demonstrated that blood distilled in the arteries could not be absorbed by the veins, engulfing and swelling the muscles. This he did by placing a man on a table and then weighing his outstretched leg. If blood were the cause of muscular contraction, as it is in a hardened penis, the leg should weigh more: "I remember having laid a naked man on a table. The table was supported on the edge of a prismatic piece of wood corresponding to the middle of the buttock where the centre of gravity of this man lay. Then the calf and quadriceps muscles were contracted, thus swelling, to stretch the legs. This should have disturbed the equilibrium, giving advantage to the feet at the expense of the head, by moving from the whole body to the legs the great quantity of blood required to swell the distal muscles. However this did not occur."[91] This experiment gives an idea of Borelli's hands-on, no nonsense approach, quite opposed to Descartes's philosophical, analogical method.

That God spoke in the language of geometry was patent to Borelli, who attributed errors in physiology to the ignorance of simple anatomists and ignorant philosophers who "are not very keen on understanding the language which the Author of Nature uses to describe his concepts in this sensible book." When asked what God did, Plato answered: "God exerts geometry." Notes Borelli:

> Animals are bodies and their vital operations are either movements or actions which require movements. But bodies and movements are the subject of Mathematics. Such a scientific approach is exactly Geometry. Similarly, the operations of animals are carried out using instruments and mechanical means such as scales, levers, pulleys, winding-drums, nails, spirals, etc. Thus it is true that, in building the organs of animals, God exerts geometry. To understand them we need geometry which is the unique and appropriate science to enable one to read and understand the divine book written on animals.[92]

This passage from his introduction indicates the exact moment and fashion in which the human body became assimilable to a material machine. Borelli himself is fully aware of the novel and momentous nature of his approach: "While many ancient and modern authors have attempted to tackle the difficult physiology of movements in animals," he writes, "nobody has succeeded so far in confirming or solving these problems by using demonstrations based on Mechanics."[93]

Before entering into particulars, Borelli—a man of obvious piety who "never allowed that a small picture of the Holy Virgin be removed from his bed" and who "was often seen kneeling and praying"[94]—makes clear that "everybody agrees that the principal and the effective cause of movement of animals is the soul. When dead, i.e., when the soul stops working, the animal machine remains inert and immobile."[95] However, the soul is unable to move the animal parts by itself. It needs an instrument which is called *will* or *motive faculty* and is supposed to reside in the animal spirits.

Borelli begins by proving that the motive faculty, the order to move, is transmitted from the soul to the muscle through the nerve. His innovation consists in asking: "But what is actually transmitted to the muscle by the nerves? Is it an immaterial faculty or a gas or some liquid or some movement or an impulse or anything else? How does it overcome the resistance of considerable weights?"[96] His answer is that the motive faculty actually resides in a *succo nerveo* or nervous juice. Through a slight shaking movement or impulse at either end of the nerves that run from the brain throughout the body, the nervous juice transmits an impulse which, in an outward flowing direction, communicates the orders of the will to the members, causing movement. In the opposite, inward flowing direction, the nervous juice transmits sense impressions to the brain.

By dissecting and observing the structure of the nerves Borelli had found that they are made of a spongy material that both accepts and absorbs wet nutrients as well as secreting some juice. The instantaneous contraction of muscles, he believes, is due to a fermentation or ebullition similar to when a concentrated spirit of vitriol is poured into oil of tartar, or any combination of acid with alkaline salt that suddenly boils by fermentation. Animal spirits, he goes on to say, are unquestionably fluid, very delicate, extremely pure, and mobile substances.[97]

They can be subdivided into motive spirits, which convey voluntary movements, and sensitive spirits, the vehicle of sensation. These are both noble, bitter, sulphurous, saline, and very active like ethanol or spirit of wine. The same noble spirituous juice is produced in the testicles and is the cause of generation. Another kind of *succeo nerveo* is the basis of nutrition in the body; it is sweet and soporific in nature.[98] In a word, fermentation and ebullition arise from a chemical combination of nervous juices, which are extremely volatile, with the alkaline nature of blood; the mechanism of effervescence is no different from that of common fermentation.[99] A very slight shaking provokes or stings the nerve endings and titillates the nerve.

Anticipating the discovery of electricity as the transmitter of messages from the brain through the nerves and hence to the muscles (not demonstrated until 1929, by German psychiatrist Hans Berger), Borelli compares the effervescent explosion of combined chemicals with the sparks that break out of the structure of stone when hit by steel. The mobility of particles that put in motion muscles is due to the innate mobility of the elements when they are freed from their jails in the material blood.[100] Why one muscle or another is thus stimulated, he confesses to not understand. The best explanation he can come up with is that the animal learns through trial and error, confirmed by habit, to move one muscle or another by an act of will.[101]

The final piece to the puzzle of animal movement regards the regularity of involuntary motions, such as the beating of the heart and pulse. In direct refutation of the Cartesian method of reasoning by analogy, Borelli considers whether "it is possible to investigate the properties of natural things in some way by using the knowledge of artificial things": for example, "an automaton seems to present some resemblance with animals since both are mobile organic bodies which comply with the laws of mechanics and both are moved by natural faculties." But if a natural body were like an automaton, reasons Borelli, it would be subject only to regular motions, and secondly, the presence of air mixed in the blood makes the functioning of organic machines quite different from that of inorganic machines. To explain these automatic movements Borelli posits an oscillatory motion put into action by minute particles of air that he characterizes as "aerial small machines," which mix with the blood and are compressed by the external forces of the surrounding viscera, vessels, flesh, and skin. Being spiral machines, they "spontaneously resile like springs, as appears in a pneumatic blunderbuss."[102]

The other reason the human body cannot be compared to an automaton is because these regular motions are affected by the passions. Proposition 79 posits that "the movement of the heart can result from an organic necessity, as an automaton is moved." Borelli is quick to deny this possibility, since the movement of the heart is not similar to the movement of clouds by the wind or the force of weights and cog wheels that power a clock. Rather, there is an innate, moral sense in human beings that somehow affects the involuntary movements of the body.

The motive principle can be traced to the workings of good and evil as they affect the conscience. For example, the pulsing of the heart speeds up or slows down depending on our emotional state, "whether some unexpected good happens or some terrible evil is imminent." Thus we can say more generally that the changes in the movement of the heart result "from fear and belief which are the cognitive faculties of the soul" and not from "some unknown organic necessity, like in automatons."[103] In other words, no matter how machinelike the material functionings of the human body appear to be, the passions of the soul are an essential component to describing *de motu animalium*. As Borelli's contemporary Carlo Giovanni observes, "A few examples of this kind show how happily he combined sciences and Catholic piety."[104]

The implications of the Borellian model of physiology for therapeutic medicine constitute a rejection of Galenic treatment of humors, purgings, blood lettings, and chemical drugs. According to Borelli, pain and disease are caused by bitterness in nervous juices that irritate the nerves and heart. This bitterness arises from a small amount of ferment in the blood that is difficult to remove. In fact, since movement is at the heart of the circulation of nervous juices, violent commotions such as fear or anger are more likely to cure a fever than are changes in temperature of the blood or phlebotomy. Most of the time the organism rights itself by its own means.

CONCLUSION

The outcome of Borelli's successful application of Galilean physics, mechanics, and hydraulics to the functioning of the human body was the creation of new "iatromechanical" and "iatrochemical" schools of medicine in Italy during the second half of the seventeenth century. Iatromechanics sought to explain the movements and illnesses of an organism according the laws of mechanics, statics, and hydraulics. Iatrochemistry sought to explain the internal functionings of the machine body in terms of chemical reactions and processes. From that time on, European culture reconceptualized its body as a machine made entirely of corpuscular particles, bits of matter in motion, differing only in shape, movement, and size. The parts of this new mechanical body that failed to fit in with the machine model—the passions of the soul, like melancholy and hypochondria—were pushed into the realm of the "psyche," where to this day they still reside somewhat awkwardly in the "unconscious," tended by psychiatrists and psychologists.

In the seventeenth and early eighteenth centuries, when it came to the question of what mediated between the immortal, conscious part of man and his mortal, material vessel, the focus of the question centered on *de motu animalium*. Descartes endowed his *homme* with a strictly material being, yet he explained the motive faculty in terms of heat without light, firing and transforming the spirits in the heart, transmitted to the brain through tubal nerve fibers which then pulled at the various muscles. Borelli rejected this Aristotelian preeminence of the heart. Sensation and desire, passions, fear, and pleasure are ultimately what move the animal for Borelli. He imagined the instrument of the motive faculty in the coursing of a *succeo nerveo* through the nerve sponges, sparking effervescence and ebullition, acids and alkalines combusting in the blood. Nerve stimulation and irritation accounted for both movement and disease. The movement of the heart could never result from blind mechanical necessity for an Italian physician; this would be tantamount to denying the existence of the soul.

Alessandro Dini, writing on natural philosophy, medicine, and religion in the time of Neapolitan physician Lucantonio Porzio (1639–1724), briefly frames the question of the *vis motiva* within a European context and shows that the Italians, beginning with Borelli, confronted this controversial question within an ontology that recognized the unity and homogeneity of material substance.[105] Chemical substances retain a qualitative as well as quantitative distinction no matter how small and homogenous the extended particles may be. For the Neapolitan physicians, human awareness and perception were intimately tied to sense experience. The body was not a mechanical vessel for an entirely different spiritual entity; it was coursed through with nervous juices, a spirituous substance neither wholly material nor wholly invisible.

In Naples, forward-thinking intellectuals, lawyers, doctors, and natural scientists had been working hard to show that atomism did not necessarily imply atheism. It was not necessary in Naples to deny a sensitive soul in order to entertain corpuscularism. Nor was it necessary to know with certainty the truth of things. As we shall see in the next chapter, the probable, the hypothetical, the experimental, the instrumental were all more prized in the medical community of Naples.

FIVE

Vico's Monstrous Body

Along with anthropologist Mary Douglas, I believe that attitudes toward the body are directly related to attitudes toward the body politic: they are direct expressions of each other.[1] Following this methodological principle, it makes sense to regard physiology and political science as reflections of each other.[2] This is borne out by the fact that in the early modern period "Physician and Philosopher" is a common epithet used by natural scientists. For Campanella, Descartes, Hobbes, Settala, Di Capoa, and Porzio, to name a few, philosophy of man begins with the physiological basis of perception and extends to forms of social organization. There was no disciplinary split in the continuum of reflection on humankind.

By examining theories in the late seventeenth century regarding health and disease I hope to shed light on how the body was viewed at a time when metaphors of machines were beginning to dominate conceptions of the body, and, thus, to understand exactly how the machine-monster became interiorized into our political notions of humanity. My line of thinking expressed in the simplest of terms goes something like this: If machines are monstrous, and the human body is a machine, then human bodies must be monstrous, too; and so, by extension, is the body politic.

This thought process is beautifully illustrated in Hobbes's *Leviathan* (1651). The political philosopher starts out with an extended metaphor based on the idea that machines are artificial animals created by man, in the same way that man is a machine created by God. His analogy between automatons and human beings is based on a mechanical presupposition regarding human physiology: "For what is the heart but a *spring*; and the *nerves*, but so many *strings*; and the *joints*, but so many *wheels*, giving motion to the whole body, such as was intended by the artificer?" Men, Hobbes argued, are really machines moved by two basic motions—the desire for power and the fear of death. In a mocking parody of his Creator's creation, man creates a being in his own image: the state, figured by Hobbes as an artificial man.[3]

The analogy presented in the preface to *Leviathan* is extended throughout Hobbes's treatise and it structures the analysis of the parts of the body politic: "*sovereignty* is an artificial *soul*, as giving life and motion to the whole body; the *magistrates,* and other officers of judicature and execution, artificial *joints;* reward and punishment, by which fastened to the seat of the sovereignty every joint and member is moved to perform his duty are the *nerves*" and so forth.[4]

Hobbes also likens negative social factors to diseases.[5] Just as all the parts of an artificial man move together to produce a semblance of life, the parts of the body politic must also function harmoniously for the state to be healthy: concord is like health; sedition is a sickness; and civil war is tantamount to death.[6] When pressed for an apt analogy for mixed forms of government, the example that comes to his mind is the Lazarus brothers whom we met up with in chapter 2 on their tour of Europe: "To what disease in the natural body of man, I may exactly compare this irregularity of a commonwealth, I know not. But I have seen a man, that had another man growing out of his side, with a head, arms, breast, and stomach, of his own; if he had another man growing out of his other side, the comparison might then have been exact."[7]

All this makes sense as a convenient analogy with which to describe the body politic, but the question remains: Why is the state, this artificial man, also depicted as a monster or a Leviathan? The answer lies once again in our definition of monstrosity as inert matter moving of its own accord.

When Hobbes returned on horseback from his visit to Galileo in 1635, writes Richard Peters in his introduction to *Leviathan,* the whole world seemed to be full of moving things. Excited by the new mechanics, excited by the discovery that motion, not rest, was the natural state of bodies, Hobbes constructed his entire epistemology based on the movements of vital and animal spirits. If the state is an artificial creation of man— an automaton made in the image of its artful creator—it is fitting according to what we have understood about the source of motion and its connection to monstrosity that it be named as the mythic sea beast from the Bible.

The biblical Leviathan is "the fleeing serpent," "the coiled serpent," most often "the dragon that is in the sea." In all cases, Leviathan symbolizes the forces of primeval chaos, an anti-creation myth juxtaposed with Genesis. The sea monster is representative of all mythological monsters, symbols of God's omnipotence. The irony resides in the fact that Hobbes's Leviathan is to man as man is to God. Man is God's creation, the climax of

God's creative activity, while Leviathan is man's creation, the climax of man's creative activity. But what man creates in this parody of God's fiat is a monster, a symbol of man's botched creative powers and essential imperfection.

Hobbes's metaphor mixes up and conflates two attributes of monstrosity that we have identified thus far: the state is a machine, and the state is an organic being prey to deformations due to the inharmonious functioning of its components. It seems to matter not whether Leviathan is a monster or a machine, really. If the human body is but a sophisticated automaton, perpetually in motion, buffeted by the winds of desire and fear, what else could it create by banding together with other such bodies, than a superautomaton? Almost imperceptibly, this phantasm takes on monstrous attributes, not because it is an automaton but because it is subject to imperfection. Monstrosity creeps into Hobbes's metaphors when it is a matter of something going wrong.

The Italian machine-man differed from the Cartesian and Hobbesian automaton in that no radical separation could divide matter and spirit in the Italian *macchina del corpo*. As we shall see, thanks to the epistemologically and spiritually united character of Neapolitan physiology, the nature of man retained an intimate correspondence between physical and moral forces, similar to the precepts of physiognomy. The monster in the body was still a monster in the soul when it came to the Italian civic philosophy of man because the source of movement in the human body could never be mechanically explained away.

Since the time of Galen physicians had reasoned that, by definition, that which is living is endowed with self-movement; therefore, there must be a portion of matter to which movement is inherent and which is the cause of all movements that take place in the organism.[8] When placed in the moral sphere and in the world of becoming as manifested in the history of nations, the physiological questions of movement, self-movement, or passive movement became reframed in terms of "Are we moved by will or fate? Are we human or brute? Are we like lesser animals and inorganic beings who are passively moved according to blind, mechanical necessity? Or are we self-moving, rational beings, endowed with particles of divine will, responding to the call of natural justice that is inherent in us?" The motive faculty served as a crucial concept that linked the issue of the soul and its passions with the issue of humanity's movement through history.

Giambattista Vico solves this tricky problem by transforming the notion of *conatus,* a term used in the physical sciences to describe the principle of motion, into a metaphysical concept that serves as the intermediary between matter and spirit and likewise between human will and Divine Will. Vico's *conatus* is the principle of motion in matter as it is the principle of motion in the spiritual movement of humanization, directed by the seemingly irrational methods of Divine Providence and played out in the courses and recourses of history. When the metaphysical nature of *conatus* is denied, humanization goes awry and monstrosity is the result, both in the individual and in the state. The need to conceive of such an intermediary was necessitated by the growing credibility of materialist philosophies championed by so-called Cartesians, Gassendists, and Epicurean atomists in the Naples and Europe of his day.

Vico saw clearly that the mechanistic philosophy could lead to a denial of human imagination and free will. To propose, as Descartes did, perfectibility of human knowledge through progressively more accurate quantification of the laws of nature was tantamount in Vico's eyes to denying the mystery of God's truth. We can know only what we make, says Vico, our own history and cultural institutions, whereas the physical world, having been created by God, is only knowable by its Maker: "God knows all things because in Himself He contains the elements with which He puts all things together. Man, on the other hand, strives to know these things by a process of division. Thus, human science is seen to be a kind of anatomy of nature's works."[9] The human mind is like a surgeon, who can only discover "the location, structure, and function of the bodily parts" by dissecting an inanimate corpse. "Consequently, the functions of these same parts can no longer be explored. . . . because of the solidification of the liquids, the cessation of motion, and the cutting [in the autopsy]."

We are the anatomists of nature; we know by cutting up the objects of our inquiry. In doing so, we alter the world we are investigating. There is a clear limit to our knowledge of that which God created. "For, by way of an enlightening example, human science has dissected man into body and mind, and mind into intellect and will. And from body it has picked out or, as men say, abstracted figure and motion, and from these, as well as from all other things, it has extracted being and unity." The way to true knowledge lies in a contrary "top to bottom" approach: the way to the physical world begins in the vestibule of metaphysics.[10]

The basis of Vico's polemic with Cartesianism was, thus, his belief that materialist philosophies end either in the deterministic rule of mechanism or in the fatalistic rule of chance. In his view, both these outlooks tended to the kind of skepticism or atheism characteristic of ironic consciousness. Ultimately, he believed, materialist thought would usher in a return to a new kind of barbarism, characterized by excessive self-reflection. In Vico's age of irony, that of advanced secular culture, when man sees only man in the workings of nature, no law can stand above him. Like the Lucretian vision of life in a state of nature described in book 5 of *De natura rerum* or as in Hobbes's perpetual war of all against all, humans will regress to bestial-like solitude amidst the crush of their throngings cities.

Vico constructs his theory of human history, therefore, as a bulwark against the approaching return to monstrosity heralded by the age of mechanization. When humans turn their body and their world into a machine, the result is a forgetting of our animal origins and thus a forgetting of our precarious struggle to achieve humanity. This struggle is reflected in Vico's narrative about the primitive, bestial giants who wandered the earth after the Flood and who only arrived at normal dimensions through centuries of ritual observances as they developed and internalized a sense of natural justice. This fantastic element in Vico's narrative—babies abandoned by their promiscuous mothers and grown to huge proportions from wallowing in nitrous salts—might strike postmodern readers as a charming, baroque conceit. It is actually consonant with the facts as explained by the physiological sciences of his day, which portrayed humanization as a process of assuming and maintaining a well-proportioned, well-tempered body. Vico's vision of human development and deformation is thus a comprehensive theory regarding the monster within.

It is unfortunate that Vico's treatise on physical medicine, *On the Equilibrium of the Living Body*, has been lost. Still, enough material remains scattered throughout his writings to piece together a convincingly vivid image of his physiology. H. P. Adams notes that "[Vico's] studies in physical science were closely connected with his work as a jurist. They afforded the great example of inductive method."[11] The fact is that metaphysics and physics are so coherently wedded in Vico's philosophy that one cannot consider one without being informed about the other. Since Vico conceives monstrosity in both the body and body politic as moral and corporeal deviation from a state of natural justice, the only way to understand the nature of that particular early modern Neapolitan body that serves as Vico's protagonist in the *New Science* is by knowing the physiology upon which its figuration is based.

Vico's familiarity with contemporary physiology is not surprising given the preeminence of the natural sciences in the Naples of his day. For over fifty years, from the last third of the seventeenth century to the first quarter of the eighteenth, Naples flourished as the most important intellectual center in Italy, and perhaps even Europe, for the natural sciences. As Max Fisch observed, "It was the [Academy of] the Investigators who had made it so."[12] Of the three academies that continued the research interests of the Investigators, "Vico was a member of the first two and in touch with the third." Vico participated in its last incarnation between the years 1683 and 1697.[13] And although he was not a member of the later

Academy of Sciences, "it was perhaps its existence under the ostensible patronage of Charles of Bourbon which led him to think in 1736 or 1737 of dedicating to Charles his work 'on physical medicine'—probably a recasting of his *On the Equilibrium of the Living Body,* which remained unpublished until the end of the century, and of which no copy can now be traced."[14]

The leading lights of the Investigators were also Vico's friends and intellectual mentors. The names of the physician founders of the Academy, Lionardo Di Capoa and Tommaso Cornelio, appear repeatedly in Vico's *Autobiography* as models for the young man's intellectual formation. Di Capoa and Cornelio had trained a generation of younger men like Vico's friend Tommaso Donzelli, who revised and enlarged his father Giuseppe's major work *Teatro farmaceutico dogmatico e spagirico* (1667).[15]

The story of the arrival of Cartesianism to Naples in 1649, thanks to physician Tommaso Cornelio, along with its reception and modifications in line with the indigenous scientific culture has been told in detail by Max Fisch, Nicola Badaloni, and Alessandro Dini. Vico himself gives a partial account of the vast and fickle enthusiasms provoked by Gassendism and Cartesianism in his *Autobiography.* Vico's solution to the breach between matter and spirit—the transformation of *conatus* into a metaphysical principle—went directly against the current of Cartesian models. It is hard to appreciate the originality of his solution without being aware of the scientific culture to which he was responding. By studying the events and issues that fueled medical controversies in Naples during Vico's lifetime, we may arrive at a clearer understanding of the human body, seat of struggle between appetites and reason, as it was conceived by Vico in the *New Science.*

The process of hominization and humanization takes place most literally in the flesh, the blood, the nerves, and the soul, guided by Divine Providence through the agency of *conatus.* Just as the blood circulates, so history courses and recourses in the spiritual movements of providence as seen in the material world of human customs. This is the way the physiological question regarding the motions of the animal body, *de motu animalium,* was reframed as a philosophical question regarding the motions of the body politic. This is what Vico did in his *De Antiquissima* and probably in his lost medical treatise *De Aequilibrio* as well. I do not think it inapt to credit Vico with being "the physician of history" as much as he is the "philosopher of history." This part of my study is thus offered both as a meditation on monstrosity during the birth of Enlightenment thinking and as a modest contribution to the vast and expert field of Vico scholarship.

In the last chapter we briefly described the metaphysical entities called "substantial forms" that, according to church scientists, mediated between matter and spirit. The philosophical explanation for the exact mirroring between immaterial soul and material body that formed the basis for the science of physiognomy is that the soul is a substantial form of the body. The Aristotelian principle that it is the proper arrangement of matter that gives rise to living beings forms a basis for this correspondence, so much so that "all living bodies are material portraits of their souls." As Monsignor Ingegneri, Bishop of Capo d'Istria, goes on to explain, while the soul (anima) is linked in some mysterious and intimate way with its corporeal counterpart, the intellect (animus) remains immaterial and incorruptible. Why is it necessary for the body to perfectly mirror the soul? So that it "may serve it": "And this is also because the soul (being the substantial form of the body) is not given any of its own properties, except for providing the vital principle, so it is necessary that to every determinate sensitive intrinsic property corresponds a determinate shape & sensible exterior arrangement of the members, adapted to those vital principles and to those arrangements that bear the soul into the body."[16]

Substantial forms, entities that divided and modified matter in various ways, thus determining the essence or nature of things, were conceived by the Peripatetics and stubbornly defended by the Jesuits well into the early eighteenth century. They were the only entities that linked physics with metaphysics, thus providing the last remaining philosophical bulwark for the church to attack new doctrines of materialism and atheism.[17] The Aristotelian scholastics further asserted that the qualities, powers, or virtues of bodies which determine their mode of operation and which produce sensations in sentient beings were additional separate entities called "species," distinct from corporeal substance. Whether these species were material, spiritual, or intentional remained a mystery.

Among medical practitioners, the issue of defining what belonged to the body and what belonged to the immaterial soul, and what the nature of interaction between the two was naturally assumed primary importance. How was a doctor expected to treat an illness without understanding its physical and spiritual causes? Therapeutic strategies were determined by such philosophical entities as substantial forms and species. Following the old doctrine, medications were classified according to their "elementary qualities" (the four Galenic humors). They were further distinguished by various qualities of "specific entities": purgatives, astringents, soporifics, sudorifics, desiccants, and so forth. Since illness was conceived as corruption of the humors, determined by imbalance of their elementary qualities, blood letting and purging were logical strategies meant to remove infected or defective humors from the organism.

In late-seventeenth-century Naples the doctrine of substantial forms assumed capital importance. This is because proponents of atomism, by rejecting any such metaphysical entities, were being charged with atheism by the authorities of the Spanish Inquisition. Between 1688 and 1697 readers of Lucretius were sent to trial, tortured, and imprisoned. Alessandro Dini describes a solemn ceremony of abjuration on the part of Carlo Rosito and Giovanni De Magistris in 1693 that links beliefs in atomism—that is to say, in nature operating as an autonomous force of order in the world of things—with denial of the Pope's temporal powers and the immortality of the soul, along with other heretical positions such as denying the sinfulness of fornication or incest.

This frontal attack on the new culture was ably defended by a number of prominent physicians, jurisconsults, and philosophers of nature, including Francesco D'Andrea. D'Andrea sought to show that there was no necessary connection between atomism and heresy and, more specifically, that the adoption of Aristotelian physics into the Christian system of ideas had no valid justification in any case. Perfectly aware that what distinguished atomism from Aristotelianism was a rejection of substantial forms, he was convinced that atomistic theories were equally suited to contemplation of the divine nature of man. D'Andrea writes:

> That which unites the Atomists, who only consider things according to what they really are, is that they do not admit in nature other entities (speaking of corporeal things) except matter alone divided into minute particles—which, as they vary in arrangement, size, shape they believe to be sufficient in forming all the various natures or properties, or may we say also essences of things, because of which one differs from the other, without having recourse to those celebrated substantial forms which the world is sick of by now, it seems to me—in the same way as all the virtues or powers or energies or whatever word we want to call them by, with which we see natural bodies made into many, which our Peripatetics want likewise to be so many entities distinct from matter, and their substantial forms from those.[18]

Other defenders of the new school of philosophy pointed to the undeniable antiquity of atomism and its links with Platonism and Pythagoreanism, both of which believed in the immanent workings of spirit or a world soul in the realm of material things. This reconciliation of (Neo)Platonism with atomism, thus rendering the latter acceptable to

Christian doctrine, was typical of Neapolitan solutions to what they considered to be an unacceptable, radical separation between spirit and body characteristic of French Cartesianism.

Other, less abstruse issues and writings rose out of equally pressing pragmatic concerns. The occasion for Lionardo Di Capoa's work, *Parere* or "Opinion," published in Naples in 1681 (and followed by a second volume prompted by Queen Christina of Sweden on the uncertainty of medications) was yet another attack by government agencies directed against practitioners of the new "spagyric" or chemical medicine. Max Fisch, in his article "The Academy of the Investigators," describes the circumstances:

> Ottavio Caracciolo di Forino, a favourite of the viceroy, had died in the full vigour of manhood, and his death was attributed to a dose of "antimony ill-prepared and inopportunely prescribed." His physician, Antonio Cappella, though a Galenist, was a dabbler in chemistry. . . . Perhaps at the instigation of the protomedico, the viceroy on 26 September 1678 laid before the Collateral Council this and other "unhappy Accidents that had befaln some sick persons, and for which the Chimical Medicines were accused." The Council asked the protomedico to appoint a commission of leading physicians to devise measures "for the putting of some stop to the abuses and errors daily committed in the practice of physick."[19]

Di Capoa was one of those leading physicians asked to comment, and when he did the literary quality of his opinions so impressed his friends that he was urged to make them available to the wider public, resulting in the most famous of all the productions of the Investigators. His *Parere* traces the origins and development of medicine in order to demonstrate the absolute uncertainty of the discipline. His strategy for opposing the intervention of legislators into medical disputes was to show that since the wisest and most able of physicians were unable to establish firm laws regarding "the practice of physick," the people and the magistrates could hardly expect to do so.

The author develops this theme by multiple demonstrations of the epistemological, philosophical, and practical controversies that have led to the absolute uncertainty of medical doctrines and therapies. For example, how can one establish the certainty of one sense by the authority of another? And if the senses can deceive, how can reason, which derives from the senses, be an arbiter of truth?[20] If reasoning is an unsound basis for medical truth, experience is no better: "On the contrary, we shall show it to be even more uncertain and doubtful." In his clever, baroque prose Di Capoa goes on to express tolerance for probable knowledge, calling to mind Vico's division of disciplines into *verum* and *certum*: "Because it surely follows then that not even reason accompanied by experience, no matter how valid, is able to render medicine certain and sure; since verisimilitude heaped upon verisimilitude, and non-certain upon non-certain, no matter how many long arguments and proofs are added to them, will never give rise to any thing that is certain and incontestable."[21]

Di Capoa paints a vivid picture of the bitter sectarianism that was dividing the city of Naples. He issues a plea to listen dispassionately to the free-thinking physicians who have turned their backs on the authority of the ancients: "And they are all those who want to follow in the footsteps of no modern or ancient, no Greek or Latin or Barbarous writer, no matter who he is, nor tie themselves to the sentiments of others, but entirely free and unbound they roam with quickened flight, skimming the vast King-doms of nature. Thus, believing in nothing from others except what comes to them through their own senses, or attested to by the most certain experiment, they wish to espy, to penetrate and to subtly examine everything with a curious eye."[22] In praise of these modern philosophers, Di Capoa names a veritable *Who's Who* of early modern scientists, lump-ing Descartes, Tycho Brahe, Copernicus, Kepler, Montaigne, Borelli, Redi, Gassendi, and Galileo all into the same school.[23] He also provides a his-tory of medical practices in Naples and suggestions for improving the training of doctors.

The importance and distinctiveness of the Neapolitan medical tradition cannot be overemphasized. Di Capoa affirms that in all branches of medi-cine—diet, surgery, and use of medications—the Neapolitans always fol-lowed their own path, one that diverged considerably from the two Greek masters. In regard to medications Di Capoa notes that Neapolitans use drugs whose names, even, were unknown to Hippocrates and Galen.[24] In surgery he traces a line of innovation from Marco Aurelio Severino (1580–1656), who wrote the first comprehensive treatise on comparative anatomy and the first textbook of surgical pathology. Fisch quips that "[Severino's] fame as a teacher, it was said with pardonable exaggeration, emptied the medical school at Padua,"[25] but at the time Di Capoa writes, the chair of anatomy in Naples itself had remained empty ever since Severino's departure, either because the surgeon considered the discipline of little worth, or because the course of study for a medical degree did not require anatomy.[26]

Severino, who restored the use of knife and cautery and practiced re-frigeration anesthesia, saw his professional compatriots as weak and effeminate in their reluctance to apply "Iron and Fire." His own virile philosophy of surgery is based on the rhetorical figures of synthesis, dieresis, and exairesis: it is "nothing but a particular manual Dexterity either to handle & treat the parts of the body that are ill, Or to take away from the body things that are foreign & harmful: Either to heal the members that are separated & disunited one from the other, reuniting them together, or if they are joined one to the other contrary to nature, by separating them."[27] Extractive surgery, also referred to by Severino as encope, involves "cutting away entirely whatever corrupts little by little, or whatever is entirely dead & corrupted; or whatever is useless & superfluous, or whatever has formed on the body contrary to Nature, or whatever has attained excessive largeness."

This rhetorical understanding of surgical procedures is in line with Severino's aim to "bring to light a Surgical body" or a constitution of the art that he calls "efficient medicine."[28] Deformation is viewed as bad corporeal form susceptible to correction by means of the knife. Whatever grows contrary to nature, whatever is superfluous or disproportionate is indicative of corruption and possibly death. The solution is an efficient one: join together, pull apart, or cut off. The only prognostication that a doctor should attempt on the basis of unnatural bodies is that of harsh medical truths regarding the future. Unlike physiognomists, the new surgeon is a hard-nosed realist who bases his diagnosis on efficiency, not analogy. Departure from the norms of beauty indicate nothing in the moral dimension except perhaps cowardice for those who were unwilling to submit to the painful extractive procedure. Doctors who seek to avoid efficient solutions to illness by resorting to diet or drugs are nothing but deceptive, effeminate sirens, birds of evil augury, he mocks. Such, apparently, were the majority of Neapolitan physicians at the time that Severino wrote.[29]

Now this is not the view of Borelli, as we have read, nor is it that of Di Capoa, who believes that a physician should be knowledgeable about the time of birth, the setting of the stars, the variety of climates, "and other similar things necessary to the profession of medicine." A medical doctor should have knowledge of anatomy; he should investigate the nature and economy of the human body, the causes of illnesses, the virtues and operations of medications, including a knowledge of chemistry; and "nobody doubts that he must be well learned in natural philosophy." In addition, ethics or moral philosophy should also be his purview, since "how could the doctor heal with valid medications those who are ill in the body, if first he does not remove the illnesses of the soul? Since all the illnesses of the body, as a first and principal cause, are usually most often born from some passion of the soul."[30]

A good foundation in herbal remedies is so fundamental that Di Capoa makes a plea to restore a garden of simples like the one cultivated by the celebrated apothecarist, "the most discerning Giuseppe Donzelli, who in simples had few equals." Donzelli had so many herbs in his shop, famous for his special theriacs, writes Di Capoa, "that he could make six or seven solutions in one day."[31] His son Tommaso, friend to Vico and editor of his father's *Teatro farmaceutico*, also edited and wrote a preface to Di Capoa's posthumous second volume. He notes with pride that this suggestion was taken to heart by the fathers of the city. Tommaso himself was responsible for stocking and caring for the plants.

Di Capoa also refers to Sebastiano Bartoli, "who used to place among the most important advantages [a man could have], to enjoy forever and make use of the greatest freedom in philosophizing, with which he achieved the undertaking of a novel system of medicine."[32] Bartoli was a protagonist in the events that brought the quarrel of the ancients and moderns to a head and culminated in the instauration of the Academy of Investigators. In 1663 the young physician resolved to publish his attack against the old medicine in the form of a series of "paradoxical exercitations," along with his "system of microcosmic astronomy" describing his new medicine. The book was instantly branded blasphemous at the instigation of the protomedico, and nearly all the copies were destroyed.

The same year (1663) an epidemic of "malign fevers" erupted in the region of Naples. The ancients attributed the cause to flax and hemp that had been left, due to heavy rains, in the nearby Lake Agnano. Their theory was that the water and air were corrupted from the rotting hemp. The moderns rejected their position and called for extended investigation. In the face of the upcoming controversy they organized themselves into a formal academy and called themselves the "Investiganti." As Fisch observes, in the eyes of the regular medical profession and the general public they were not inquirers but partisans; they were ridiculed as "chemists" or "spagyrists."

In 1666 a rival academy was organized and called itself the "Discordanti." "The sessions of the Discordanti were devoted to confrontations of the Galenic and modern doctrines in medicine, usually to the disadvantage of the latter. So hot was the rivalry between the two academies that the latest charges and countercharges became the news of the day in Naples."[33] Successive events prompted Bartoli, Di Capoa, and Cornelio to ink many a page in defense of the new philosophy.

Tommaso Cornelio (1614–84) has been mentioned as the bearer of Cartesianism to Naples in 1649. Matteo Barbieri, writing in 1778, also credits him with being the first person to apply knowledge of physics to medical uses in Naples.[34] Cornelio studied mathematics and physics in Rome and Naples, "especially that part of Physics that applied to the Human Body is called *Physiology*." Traveling from Rome to Florence and Bologna, he made friends with, among others, Michelangelo Ricci, the mathematician, and with Torricelli, close friend of Galileo and inventor of the barometer. With the works of Galileo, Gassendi, Descartes, Bacon, Harvey, Boyle, and other moderns in hand, he returned to Naples, where he taught mathematics and medicine at the university. There, says Barbieri, "he sparked the spirits of the Young people to new things in Physics, and in the Economy of the Human Body."[35]

Vico refers to the prose of Cornelio's physiological essays as an example of incomparable but overwhelming Latinity, due to its "extreme purity."[36] Partly because of the plague in 1656, which carried away his surgeon friend Marco Aurelio Severino, Cornelio delayed publication of the book until 1663. The delay cost him the credit for his discoveries in physiology, claimed first by the English physicians Willis and Glisson. This was only one of his difficulties: "His own merits procured for him Accusers in Naples, who withheld mention of his work out of disbelief and who held his discoveries to be Fables," a fate common to men of innovation, comments Barbieri.[37] Interestingly enough, due to his indefatigable efforts, Cornelio, who specialized in the field of nutrition and digestion, became ill with Hypochondria, an illness of the stomach and spleen that literary men such as Vico tended to contract. He died of it in 1684.[38]

In his *Progymnasmata physica*, seven essays on physiology, Cornelio writes about his discovery that foods arrive in the stomach, where they are digested and broken down by fermentation and by heat, compressed by the motion of the stomach walls, and loosened up by solvent juices until they are reduced to chyle, the milky fluid that passes through the thoracic duct into the veins. His innovation consisted in showing that while the foods make their circuit they are converted into blood, not in any particular part of the viscera, but throughout the whole body. A summary of his essays appeared in the *Philosophical Transactions* of the Royal Society of London in 1665.

In further research, Cornelio examined the causes of life, which he saw as residing in the motion of the blood and in the respiration of air. As a follower of Cartesian materialism, he believed that air could be formed out of water, as in the case of marine animals who breathe under water, through the simple modification of particles. He departs from Descartes, however, in attributing the life force to the continued motion of the blood, "not from the *heat* of the Blood (as *Des Cartes* would have it) but the moist steams and expirations of the Heart."[39] Cornelio also sees a limit to the progress of knowledge, uncharacteristic of Cartesianism but typical, as we have seen, of the skeptical atmosphere that pervaded Neapolitan medical circles. In the third essay, *De Universitate*, the Neapolitan philosopher "seems to be in a Maze, and thinks, That the *Structure* of the *Universe* hath not been understood hitherto, nor will easily be hereafter."[40]

Writing another work, *Platonic Circumpulsion*, Cornelio advanced the theory that all movements come from air pressure and not from any "Attractive Qualities." These studies led to others on hydrostatics. With Severino dead, Cornelio dedicated one of his three essays in *The Knowledge of Air, Water, and Fire* to Giovanni Alfonso Borelli in the name of his departed friend. In it he imagines the ghosts of all the doctors and philosophers milling about the Elysian Fields and deriding "the vanity of Astrology, and the garrulousness of the Medicine of his time."[41]

Another Neapolitan medical innovator was an intimate friend of Vico. In his *Autobiography* Vico tells of long and frequent discussions on medical topics with Lucantonio Porzio (1639–1724), "which won him the latter's high esteem and intimate friendship, maintained until the death of this last Italian philosopher of the school of Galileo."[42] Alessandro Dini, in his monograph on Porzio, observes that "in reality . . . if one wanted to characterize the personality of the Neapolitan physician with an appellative, it would be more apt to define him a Cartesian."[43] No wonder, then, that Vico's unusual physiological speculations based on the mechanical medicine of the Egyptians, that of "slack and tight," caused the good Porzio to be dismayed! Vico, with characteristically ingenuous pride, tells us that "Porzio used to often say to his friends that the things meditated on by Vico, to use his own expression, had a hypnotic effect on him [*il ponevano in suggezione*]."[44] In a passage from his *Collectio phrasium* Vico tells an anecdote about his friend that gives an idea of the state of medical practice of his time: "One time, while consulting with other physicians about a patient whom he believed to be incurable and who would have been killed before his time by the remedies that the other physicians had proposed, Porzio cut short all debate by coining for himself an aphorism that they believed came from Hippocrates or Galen: 'It is better not to kill a patient than not to cure him' [Melius est non occidere quam non sanare]."[45]

Porzio, along with another physician and academician, Lucantonio Tozzi, both sought to apply physics to the human body. Porzio's contributions include *De motu corporum*, in which he uses a new method to determine the ratio between absolute and relative gravity on bodies, by which they derive their principle of movement. In the second part of the work he presents studies on hydraulics and gives plans for the construction of a number of machines.[46]

Dini describes how this theory provoked quite a debate among the academicians, giving rise to numerous publications, most notably one by Paolo Mattia Doria, another close friend and interlocutor of Vico on medical theories and the one to whom he dedicated his *Liber metaphysicus* on the ancient wisdom of the Italians.[47] Doria, says Vico, writing about himself in the third person, was "as fine a philosopher as he was a gentleman, [and] the first with whom Vico could begin to discuss metaphysics. . . . In Doria's discourse he perceived a mind that often gave forth lightning-like flashes of Platonic divinity, so that thenceforth they remained linked in a noble and faithful friendship."[48] In reading Doria's prose one hears subtle echoes of Vico's ideas. For instance, Doria's objection to Porzio's theory of animal motion was that by distinguishing a point on a sphere and then calculating the specific gravity of the body by a perpendicular line to the earth along an inclined plane, Porzio had committed the error of considering "physical things with metaphysical mind."[49]

Just as Vico saw Plato's physics—matter created out of a metaphysical principle "like a seminal spirit that forms its own egg"—as being the basis of a moral philosophy founded on an ideal or architectonic virtue or justice,[50] Doria believed that the study of geometry was a valid method "to form a man equally learned as he is well-mannered, and to make him become equally useful to himself as to the Republic."[51] As the incipit of Doria's *Dialoghi* indicates, knowledge of the true and the false could only be derived from synthetic geometry. In cultivating abstract methods for arriving at the one principle of truth and justice in the metaphysical realm, young people would be better trained to discover such principles in the moral and political realm. By searching for truth and falsehood through mathematics young people can remember that which their own minds have created: "che la sua mente ha fatto." Contrary methods, he says, "ruin and corrupt human reasoning."[52] Geometry leads to knowledge of the one truth, which provides a metaphysics. This in consequence gives the human mind an idea of Justice and of all the characteristics of the just idea, which is the key to educing true humanity out of young minds and bodies.[53]

In *On the Most Ancient Wisdom of the Italians* Vico says that Doria is the only man of his generation to have applied metaphysics to human uses, "adapting it in one way to mechanics and in another way to civil doctrine."[54] The consonance between Doria's ideas on the studies of his times and Vico's is striking. Doria, too, sees the growing trend toward mechanization as nothing less than a monstrosity. He objects bitterly that, having turned mathematics into a chimera, the modern geometers have proceeded to make of it a physics and a mechanics. The link between physics and political philosophy is a direct one: "Consequently, following the same disorganized method in the study of Philosophy, of laws, and of morals, men have in these most important Disciplines, followed so much monstrosity in various maxims."[55]

By this he means that when you derive the art of governing from physical principles you will end up with the monstrous kind of prince that

Machiavelli came up with. Vico provides a helpful gloss: by deriving instead the art of governing from the one true metaphysical principle which is God, Doria was instead able to form a prince "untouched by all the evil arts of rule with which Cornelius Tacitus and Niccolò Machiavelli endowed their prince. No teaching is more conformable than yours to Christian law, none more conducive to the prosperity of republics," says Vico to Doria.[56]

Doria, like Vico, further believed that God had created a subtle matter, or pure ether, that was endowed with a continuous circular motion (*conatus*). The shapes and movements of sensible bodies derived from the union and movement of the infinite insensible bodies, the particles of ether. Just as the movement of human body parts depends on animal spirits that act either as a motive faculty or in automatic fashion, it is due to the movement of the subtle matter that bodies accelerate uniformly down an inclined plane.[57] "You are my authority, most illustrious Paolo," writes Vico, "for this tenet that there are actions in physics and powers in metaphysics."[58] Considering the priority that both men granted to metaphysical bodies as the explanatory foundation for physical bodies, it makes sense that Doria, as Vico tells us, "was highly pleased" with his attempt to extend the metaphysical notions of ancient Italian physics to the advantage of modern medicine.

In spite of his admiration for Di Capoa's "good Tuscan prose" which he had "clothed with grace and beauty," Vico viewed the sort of skepticism he propounded as damaging to morale and to the pursuit of knowledge in general. It scatters and disperses the community of men instead of uniting it. While domesticated animals at least have the talent of gathering together in herds, the modern skeptics live like savage, grotesque beasts, fearfully divided and alone in their lairs and dens.[59] In his autobiography Vico refers to the citywide literary polemic that in 1680 had burned through the city of Naples sparked by an epigram hurled at Di Capoa by an opponent.

At that time, Vico explains, he was friend to the members of the pro-Di Capoa faction. Nevertheless, in direct reference to the damaging consequences of the *Parere*, Vico writes later of that period that

> medicine, because of the frequent revolutions in systems of physics, had declined into skepticism, and doctors had begun to take their stand on acatalepsy or the impossibility of comprehending the truth about the nature of diseases, and to "suspend judgment" or withhold assent when it came to diagnosis or application of effective remedies. Galenic medicine, which when studied in the light of Greek philosophy and in the Greek language had produced so many incomparable

doctors, had now, because of the great ignorance of its followers, fallen beneath contempt.[60]

Writing in 1726 to Padre Edoardo de Vitry on the state of culture in Naples, Vico notes that "physics are no longer put to the test, to see whether they hold up under experimentation" and "medicine, entered into skepticism, also maintains the rhetorical 'skeptical doubt' of writing."[61] From Vico's point of view it was not philosophy but rather imagination that gave rise to all the great inventions of the returned heroic times, such as the alembic (a glass vase over a flame used in alchemy), "which has brought about with spagyrics so many advances to medicine; the circulation of the blood, which has changed our sentiments toward the physics of the living body and an about-face in anatomy."[62] No enemy to Galenic medicine, nor to medical advances based on rational, empirical, or chemical approaches, Vico welcomed progress as long as it was useful to the human condition. But knowledge should be one knowledge, not divided by eclecticism: "To-day, instead . . . you might happen to be directed, for example . . . to learn physics from an Epicurean, metaphysics from a Cartesian, theoretical medicine from a Galenist, practical medicine from a chemist."[63]

The kind of skepticism put forward by Di Capoa appeared to advance only a nihilistic attitude toward the possibilities of knowing with certainty. This sort of attitude in turn would make minds more receptive to the kind of specious "clear and certain" knowledge that Renato Delle Carte offered to weak thinkers, like Epicurean materialism, suitable only for children and women. That Descartes had simply borrowed Epicurean physics, Platonic metaphysics, and outdated Aristotelian physiology and, eager for glory among professors of medicine, had passed them off as his own consistent system was evident to Vico. The *Passions de l'âme* he saw as more useful to medicine than ethics, but even that was debatable in his opinion, since (as we would now agree) "the anatomists do not find the Cartesian man in nature."[64]

Coming from a city of activist physicians engaged in cutting-edge research, the observation must have come easily to him. In spite of the medical studies he undertook as a young man[65] and despite his general interest in the physical sciences, Vico saw his vocation as the philosophy of humanity and all else as derivative to that aim. When he became aware of the "growing prestige of experimental physics," he looked into it but came to the conclusion that "profitable as he thought it for medicine and spagyric, he desired to stay far away from it, both because it added nothing to the philosophy of man and because it perforce explained things in barbarous ways."[66]

By which I understand that experimental science used a language of physical concepts unsuited to metaphysical understanding, and that it failed to take into account the fallen nature of man, deformed in his understanding, his language, and his spirit. A passage from his *Collectio phrasium* would indicate that the analogical, microcosmic medicine was just as useful to Vico for the understanding of human nature as was the empirical sort because it perceives more accurately man's true fallen nature: in the human soul, says Vico, "there coexist the wrath of a lion, the ferocity of a tiger, the shrewdness of a fox, the lustfulness of a dog or goat, the hunger of a wolf, the prudence of an elephant, and other passions in which are marked the other species of animals."[67]

Vico and Illness

How did his own fallen nature manifest itself to Vico? The question might seem frivolous. What need is there to personalize or go digging about in a philosopher's private life in order to understand his ideas? Yet the approach I am about to adopt—a reading of Vico's body as a paratext to elucidate his understanding of human physiology—is as illuminating as it is intimate. "I am in pain [wrote Vico] and yet I do not know any form of pain, and I know no limits of the soul's illness. This cognition is indefinite and, being indefinite, it is fitting for man. The idea of pain is vivid and illuminating like nothing else."[68] The more we discover of the philosopher's personal battles with illness and physical pain, the easier it is to piece together the fallen nature of human beings and how they may go about redeeming themselves from a state of monstrous bestiality into a state of heroic virtue.

The theme of physical pain linked to metaphysical suffering is, in fact, a continuous and heartfelt theme in Vico's literary corpus. He first writes about it at age twenty-four in his "Affetti di un disperato," fruit of a long convalescence at Vatolla to recover from tuberculosis. "In 1692," as H. P. Adams notes, "he reached a condition of despair in which he feared that his body would no longer answer the fierce requirements of his intellect . . . [and] he put his soul into a poem for the first and last time."[69] "The Sentiments of a Desperate One" begins with a Petrarchan lament to the poet's "bitter martyrdoms" that bear him unceasing torments. The nature of these torments is not love, though, as one would expect from the "Lasso, vi prego" opening: young Vico suffers from a defective, unnatural connection between body and soul.

His two halves are perpetually pitted against each other in bitter enmity. When the soul of any animal reaches the threshold of life, he explains in verse 43, the flaming vigor that is a property of the life force normally unites the spirit to the body with "sweet, friendly bonds." In his case, either because of adverse, cruel chance, or from lack of help from an ungenerous star, or because nature had erred off her course when he was born, she "composed me as two foes."

The young man goes on to describe how agonistic body and soul mutually torment each other in a sort of *psychosomachia*: "the mortal part of me, afflicted and tired / that now, it seems to me, is failing, / assails the soul with harsh and wearisome pains; / and my better half, who abounds with sick cures / afflicts the body with cruel plagues":

> il mio mortale infermo, afflitto e stanco,
> ch'omal par venir manco,
> strazia l'alma con pene aspre, noiose;
> e 'l mio miglior, che d'egre cure abonda,
> affligge 'l corpo con crudeli pesti.[70]

The medical conceit that Vico uses to describe the protracted war of illnesses and pharmakons hurled between body and soul is not simply a rhetorical flourish. He had always had delicate health and had almost died of tuberculosis; the constant care of bickering doctors and the uncertainty of medications and therapies must have seemed like "sick cures" indeed to the ailing adolescent.

But most notably what we read here is an intimate relationship between body and soul that diverges from the strictly Lucretian cast of the rest of the poem's natural philosophy. This iron world is coming to a close, he writes, and the fates have already been instructed to bring us to destruction, a destruction proportionate to our crimes, evil for evil, greater than the mass of evils in the ancient ages, because now we groan under the weight of new diseases. Our heavy bodies, frail, lifeless, pale, bow down while our life spans shorten and our winged lives go faster and swifter toward the grave than ever before:

> Perchè cadente omai è 'l ferreo mondo
> e son già instrutti a farci strazio i fati,
> di pari con le colpe i nostri mali
> crebber sugli altri delle prische etati
> troppo altamente, poiché sotto il pondo
> di novi morbi i gravi corpi e frali
> gemono smorti, ed a la tomba l'ali
> il viver nostro ha più presto e spedite . . .
> (lines 22–28)

It must have seemed to Vico, afflicted with tuberculosis and recovering for years in isolation, weakened and frail like an old man, that his body really wasn't up to the demands of his soul; that his soul fit his body badly and that the intense, continuing physical suffering that he experienced was some kind of fit punishment for his moral frailty. This correspondence between the physical and the moral aspects of his health is evidenced in the carefully crafted vocabulary: *morbo* means an infectious disease, a pox or a plague. But it also means a scourge, an evil, *un male morale*. *Male* itself means both sick, not feeling well, and evil or wickedness. The iron world is *cadente*, "ruined," "tumbledown," "decrepit," but it also means "an old man." The setting sun is "cadente" and a falling down building is "cadente": it means both falling into decay, growing old, and going into ruin.

These words continually draw the medical lexicon simultaneously into the historical world of time and into the moral world of affects. A weak body for a weak spirit, just as physiognomists would predict. Another passage in his *Collectio phrasium* supports the likelihood that Vico was familiar with this science and followed its precepts. From the distinguishing animal markings that surface in human passions—the wrath of a lion, the lustfulness of a goat, and so forth—it would seem that God gave all these natures to man, making him the king of the animals and the most perfect of all the creatures: "Except that when man does not dominate all these passions with prudence, the principal nature with which he was endowed, then yes, one may say that he is not only the ugliest and most ferocious of the beasts, he is the devil."[71]

In the remainder of the poem Vico goes on to describe how he manages to live with his perpetual internal struggle against physical suffering: "and when, woe is me! with many a thought and often / me against my self inside me do I hear / I have no member with which to respond to my soul / since I have no virtue that my senses awaken / except however much they make me feel / the mordant effects of their disdain and wrath. / In such a sorry and such a miserable state / go on, hope, if you may, for some rest."

> e mentre, oimè! con pensier molto e spesso
> me 'interno a sentir me contro me stesso,
> membro non ho ch'a l'anima risponda,
> poiché non ho vertù che i sensi dèsti,
> se non se 'n quanto mi si fan sentire
> gli acerbi effeti de' lor sdegni ed ire.
> In sì misero stato e sì doglioso
> va', spera, se tu puoi, qualche riposo.
> (lines 56–63)

The double entendres are significant in this passage as well: the thoughts that he both hears and feels (*sentir* also means "to perceive") are both many and *spesso*. *Spesso* means "frequent," but it also means "thick, dense, or crowded." A forest is "spesso," as is a wall or a bank of clouds and fog. While Vico's thoughts become more frequent, they also become more crowded, as animal spirits were wont to do, blocking up or suffocating his sense perception by their slow-moving density. He has no part or component with which to respond to the prickings of his soul. His virtue or power of movement is impotent, leaving his *membro* unable to respond. *Membro* is another double entendre. It could mean a limb of the body, more specifically a penis, or a member of government, a constituent. The only motion of the soul that awakens his senses are the violent commotions of anger or scorn, which were thought to produce agitation of the animal spirits. How then, he asks, can he find a state of *riposo*—yet another word that spans the physical and moral lexicons. *Riposo* could mean a relaxing state of the body or the finality of death, as in "rest in peace," or it could mean "tranquillity" of the spirit, a state of peace and justice between the warring factions.

The war between body and soul is manifested in an internal, self-reflexive dialogue: thought upon thought. Hyperbolic doubt of the Cartesian sort is precisely what brings the young Vico to ask himself what good intelligent self-awareness brings to a clever thinker: "A longing comes to me / only from a luminous part of the sky to awaken / at the foot of beeches and under the shade of laurel / the beautiful light that was partly extinguished in my poor soul / when it clothed itself in the veil by which it obscured itself: / so that up until now obstructed by another sort of stupor / it seemed to say to itself:—Wretched one! who am I?— / Woe is me! that I travail for that wish of how / I should name myself; / but I shall always call it pain and not a gift, / that those who know better what ails them suffer more."

> Mi venne sol da luminosa parte
> del cielo una vaghezza di destare
> a' piè de' faggi e poi de' lauri a l'ombra
> la bella luce che fa l'alme chiare,
> ch'a la povera mia si spense in parte
> quando se 'ndossò 'l velo onde s'adombra:
> talché, d'altro stupor finor ingombra,
> parea a se sessa dir: —Lassa! chi sono?—
> Oimè! ch'a tal desio travaglio come
> debbami dar il nome;
> ma sempre 'l chiamerò pena e non dono,
> se affligge più chi più conosce il male.
> (lines 106–17)

The rest of the poem goes on to congratulate peasants and shepherds, the primitive peoples, for their blissful state of ignorance. In this passage, too, the medical metaphors abound in puns. The principle of heat and life brings with it a longing, but *vaghezza* could also mean a vagueness, an uncertainty; it could also mean a pleasure or a delight. It is both a sensual desire and a haziness of the spirit. But the warmth and clarity of the beauteous light of spirit has up until now been obstructed or burdened with another sort of stupefaction. *Stupor* can either refer to wonder and amazement, the affect associated with *meraviglia,* or it can be a medical term indicating the state of stupefaction, when the animal spirits are so agitated and crowded that they result in statuelike symptoms. Dumbness, numbness, paralysis of the judgment are the physical symptoms of "stupor." The use of the word *travaglio* extends the medical metaphor. It can either mean "labor, toil, and drudgery," physical torment and torture, or suffering in general, as in "anguish or distress." But it is also a medical term indicating pain, as in "stomach upset" or "labor pains." The final lines indicate that the material and spiritual parts of his self suffer equally with the pain of self-consciousness. Self-knowledge is not a gift but a *pena:* both a pain and a punishment, a chagrin for the body and a chastisement for the soul.

The list of physical complaints scattered about Vico's writings is a long one. They run from a fractured skull at seven to an abscess in his throat that eventually ate away his palate. At the time he learned of this ailment in 1727, he was preparing his response to the editors of the *Acta lipsiensia.* The book had to come out quickly in order to undergo what sounds something like the early modern equivalent of chemotherapy: "Because of a gangrenous ulcer growing in his throat . . . he, although sixty years old, was persuaded by Domenico Vitolo, a most learned and experienced physician, to risk the dangerous remedy of cinnabar fumes, which, if by mischance they reach the nerves, bring about apoplexy even to the young."[72]

At eighteen years of age he was battling the tuberculosis that occasioned his nine-year sojourn in Vatolla. He was afflicted with scurvy in 1729.[73] While writing the life of Maresciallo Antonio Caraffa he was tortured with a two-year bout with "crudelissimi" spasms in his left arm due to Hypochondria.[74] And throughout his life, due to his melancholy and irritable temperament, Vico battled hypertensive nerves that often made him querulous and impatient with detractors. In his old age he describes himself as continually "tormented by spasms of pain in his thighs and legs."[75] If we are to believe any of Villarosa's account of his last days, they were excruciating, punctuated by a series of inefficacious treatments prescribed by his physician friends and colleagues at the university. His nervous system, his memory, and his frame all collapsed under the weight of years and fatigue.[76]

These physical complaints must be viewed against the background of his temperament, inherited from his parents and aggravated by the early fall off the ladder in his father's bookstore: his father was of jovial humor, while his mother was "of an exceedingly melancholic temper." The combination of opposing humors, as we have seen from physiognomic principles, led to the character of their son. Thus, as a child and up to the time of the fall Vico was high-spirited and impatient of rest: a natural justice balanced the opposing humors from his parents and resulted in a state of volatile, agitated spirits associated with liberality and boldness. Having lost consciousness for five hours as a result of the fall, deprived of both sense and motion, the little Vico subsequently "grew up with a melancholy and irritable temperament."

He was literally fallen from virtue and from wholeness: he was deformed by the growth of a tumor underneath the skin of the cranium:

> But at the age of seven, having plunged headfirst from the top of a ladder to the floor below, where he remained for all of five hours without motion and deprived of sense, and having fractured the right side of the cranium without breaking the tough skin, subsequently a large, misshapen tumor was occasioned by the fracture; because of the many and deep lancings the child lost much blood: so that the surgeon, having observed the cranium and considering the long prostration, predicted of it that either he would die of it or would have survived a fool [*stolido*].[77]

For the remainder of his life, Vico was haunted by this prediction. In 1708 he writes that "all through my life a single thought has provoked in me the greatest fear—that of being the only one to know—something that has seemed extremely dangerous to me like the one that presents the alternative of being either a god or a fool [*stolto*]."[78] Several clues help us to understand the importance of the dichotomy in Vico's mind between being a god and being a fool. The first regards the heroes of the second age, who believed that they were descended from the gods. Their primitive nature was still choleric and obstinate—"collerici e puntigliosi"[79]— but their divine origin was what separated them from their slaves, whom they believed to be the products of wicked bestial communion ("infame comunion bestiale"). Having come to them without gods, the heroes took them for beasts. As Mark Lilla observes, "The hero is only midway between man and beast, and not, as the Greeks anachronistically believed, between men and the gods."[80]

From other passages we can affirm that the beast and the fool are one and the same: "The man that is a fool is among animals the most ferocious."[81] This is clear from Vico's conviction that we are all fallen when we come into the world. It is up to us to learn how to dominate our passions to bring them into the light of natural justice:

> And if I say that each of you must search within himself in order to consider carefully his human nature, he will in truth see himself to be nothing but mind, spirit, and capacity for language. Indeed, when he analyzes his body and its functions he will judge it to be either that of a brute or in common with the brutish. From this he will note that man is thoroughly corrupted, first by the inadequacy of language, then by a mind cluttered with opinions, and finally a spirit polluted by vice. He will observe that these are the divine punishments by which the Supreme Will punished the sin of the first parent so that humankind who descended from him will become separated, scattered, and dispersed.[82]

In the *De Antiquissima,* Vico defines more exactly what a brute is. Interestingly enough, what distinguishes an animal from a man is the same as that which distinguishes a man from a machine and natural creations from unnatural ones: its source of motion. "For the Latins *brutum* signifies 'immobile'—Brutes are moved by attendant objects serving as devices. . . . For to them brute meant the same thing as immobile, yet they saw that brutes were moved. So the ancient philosophers of Italy must have held that brutes were immobile because they [were not self-moved but] were moved only by external objects serving as a mechanism, whereas men have an inner principle of motion, namely the animus, which moves freely."[83] This passage would remain esoteric and obscure had we not read first Borelli and the other physiologists of Naples. The principle of motion lies at the base of any definition of human or brute, machine or monster. Whether we are self-moved by the freedom of our spirit (animus), or whether we are by nature immobile—matter passively moved according to blind, organic necessity like machines and animals—is tantamount to asking whether we are godlike human beings or foolish beasts.

Our task in this life is to examine ourselves for evidence of the crimes we have already committed in a previous life and to compensate for that corruption. The power to discriminate, to know thyself, is the same faculty of reasoning that creates metaphors, the signifying capacity of making one thing stand for another. The power by which the human mind compares things together or distinguishes one from another is an act of discrimination that brings us closer to God: "Indeed, what may explain the fact that with one single act of perception of the eyes, we can see ugliness or deformity in things? Is it not that same power by which, for example, we can inspect all the members of the human body, compare the one with the other, order them, and see how harmoniously they are related? We can, then, determine what is fitting, what is alien, what is missing, what is best. Thus the judgments that in an instant are formulated by the mind are as many as the parts of the body (which are indeed almost infinite)."[84]

Why compare the power to discriminate the harmonious and the alien with the power to inspect the members of the human body for ugliness and deformity? Clearly, the one is an analogy of the other. Deformity of limbs is analogous to deformity of reason. It behooves us to ask what fits together properly, what is alien, what is missing, and what is best. This is the task that the fool fails to undertake. His lot is not a pretty one, since the fool, like the Desperate One, is subject to a perpetual war between his body and his soul. He lives in the dungeon of his bodily senses, besieged by passions: "The fool is always in conflict with himself, always hostile to himself"; "he has surrendered himself into a slavery under the most ruthless of tyrants."[85]

Vico draws out this theme at length in his *Second Oration*, given in 1700. Comparing the horrors of the battlefield to the self-inflicted suffering of the fool, Vico admits to speaking from personal experience. The weapon of the fool, he says, is his own unrestrained passion:

> The fool suffers that most grave of evils which Persius bequeathed to him with these eloquent words: "Let them see what virtue is and let them be tormented by its rejection." This happens because our reason is moved by the beauty of virtue, for which it has been born, and can hold both affection and passion at bay. But this is not so for the fool. The fool, like Homer's Hector, may be swept away by restless horses. Not knowing the limits of the right, "within which the just can only exist," while he tries to avoid vice he succumbs to its opposite. . . . Throughout his life he either burns with desire or shivers from fear. He either becomes consumed with pleasure or is overcome by anxiety. The soul of the fool is besieged by these passions from beneath, by these batterings and assaults. . . . He is remorseful, annoyed with himself, and troubled. He is uncommitted. He changes opinion daily, moving from extreme to extreme. . . . We can speak of him as of Plautus's Alcesimarchus: "his spirit is not where he is, it is where he is not." He is always finding cause for self-condemnation. He is always seeking excitement, never in touch with himself. He is always searching for new surroundings, new responsibilities, new ways of life, initiating new hopes even to the time of death. He is forever fleeing from himself.[86]

"O how perverse are the thoughts of the fools!" Vico remarks in another context. Swept away by restless horses, the fool appears to be constantly in motion, perpetually driven from the pursuit of one desire to another, buffeted by the winds of passion. But he is not self-moving. He has abdicated his capacity to master his motions, and thus, similar to the brutes, he becomes a passive agent, moved by the passions instead of moving through his own reason, his animus. The wise or *sapiens* is he who "through knowledge . . . separates the spirits from the concerns of the body, thus allowing him to devote himself to the better and godlike part."

The fool, deprived of his freedom of motion, is "confined in a dungeon of impenetrable darkness and surrounded by terrifying things."[87] Vico concludes his oration by linking the defectiveness of the body, its fragility of perception and judgment, with the perversity of the passions: "I believe that by now you have understood well what I am saying. The dark dungeon is our body. The wardens are opinion, falsity, and error. The guards are the senses, which are the keenest in childhood but dulled by old age and throughout life severely impaired by perverse passions. A disease of the nerves, a defect of the organs, or an intemperate desire will alter and reduce their power. What? Diverse bodily structures cover many different and even contrary natures?"[88]

The three remedies to our corrupt nature consist of wisdom "to tame the impetuousness of the fools, with prudence to lead them out of error, with virtue toward them to earn their goodwill." Those who follow the path of wisdom "are indeed men much above the rest of mankind, and if I may say, only a little less than the gods."[89] He exclaims in the *Third Oration:* if only "God Eternal had made man subservient to his own nature like all other creatures! With his will thus shackled, man would then follow the course of right reason for which he was intended. He would strive to achieve his goals in a manner of even greater consistency than that of the sun and stars tracing out their course." Throughout his life, the wise man would show forth examples of justice: temperance, fortitude, and prudence. "He would never for pleasure abandon his human nature and be transformed by some Circean potion into a brute animal intent on its own gratification."[90]

Vico stresses the constancy, stability, and regularity of motion of the *sapiens*, subject to the same movements of the earth and the celestial bodies, like that of inert matter guided inexorably by the laws of nature. Would that man, gifted with freedom of will, could stabilize the motions of his animal spirits in such a regular rhythm. We, however, are subject to the wobblings and deviations of the passions. By a strange paradox we are slaves to our own freedom of choice, which is the source of "all misfortunes, all ruin, all plagues of mankind," "the reason for all evil."[91] Fool or god, the choice is up to us. Knowing the mechanisms and the physiology of the passions is the best way to learn to understand them and dominate them. The fall off his father's ladder and ensuing fear of remaining a fool for the rest of his days take on a certain poignancy when read in the light of these passages.

Only from this perspective can we understand the unified character of Vico's program of studies as a remedy to the corruption of both body and soul. Wisdom admonishes us to embrace the whole sphere of human arts and sciences, including knowledge of things divine and things human. Natural things include physics, anatomy, and the medical arts: "Under physics I include anatomy, which is the study of the fabric of the human body, and that part of medicine which inquires into the causes of illness and which is nothing else than the physics of the diseased human body. That which provides the treatments of illnesses is indeed properly called the art of medicine and is an effective corollary of the integration of physics and anatomy, such as mechanics is the practical application of the integration of physics and mathematics."[92]

Only a complete cycle of studies can bring corrupted man into the knowledge of right. This is the basis of Vico's criticism of Di Capoa, discussed earlier. Scholarship should unite men and philosophers. Those who simply critique are not abiding by this law: "Is the physician abiding by this law or disregarding it when he contributes nothing of his own to the common well-being but rather dissects away the medicine of the ancients from that of today?"[93] Vico urges students rather to study with respect those who precede them. Descartes's investigations on the motion of bodies, the passions of the spirit, and optics come under favorable review in this passage. Just so, Galen: "To you who will be entering the practice of medicine, search Galen and learn with what elegance he assigns names to diseases, with what penetrating ability of mind he detects symptoms, and how many diagnoses he makes with accuracy. You will allow him to be the greatest of physicians."[94]

This is not to suggest that the ancients are a repository of wisdom. Vico follows his praise of Galen with an imagined conversation between a modern anatomist and a Peripatetic who refuses to believe that the nerves do not originate in the heart. The dialogue is amusing and exasperating, with the follower of the ancients refusing to perform the dissection and, once the cut is made, refusing to look. It reads very much like a number of such anecdotes in Galileo's and Di Capoa's writings in which a foolish scholastic, forced to consider the evidence of his senses, nevertheless denies it. The orator is making a plea for open-mindedness, tolerance for opposing opinions, humility when it is necessary to change them. The passage contains a number of allusions to phrases taken from Horace, Terence, and Catullus;[95] but despite its rhetorical casting, the imagined dialogue conforms so closely to similar anecdotes like the one Vico tells of Porzio, that I am convinced it came from a similar personal experience.[96]

In his *De Antiquissima* Vico reiterates that "the belief that the nerves originate in the heart has been found to be false by contemporary anatomy; for the nerves are observed to spread out through the entire body from the brain, as if from a tree trunk." In this passage he specifically refutes not the Peripatetics but Descartes, who situates the human spirit in the pineal gland. In another passage he explicitly links the question of *de motu animalium* with the most difficult metaphysical question of his age:

> This is where those thorns and brambles come from, with which the subtlest metaphysicians of our time scratch one another and get pricked in turn, when they inquire how the human mind acts on the body or the body on the mind, since only bodies can touch other bodies and be touched by them. Driven by these difficulties, they resort to a mysterious law of God as to a device, explaining that the nerves arouse the mind when they are moved by external objects and the mind pulls on the nerves when it wishes to act. Thus, they feign the human mind to be in a pineal gland, like a spider resting in the middle of its web. When a thread of the web is moved in any direction, the spider feels it. However, when the spider feels a thunderstorm coming on, the web being still, she can make all the threads move with it.[97]

But if the spirit were located in the head, asks Vico, how can you account for people with brain damage who are still able to successfully reason, feel, and move? The spirit, he goes on to say, couldn't possibly reside in a part of the body like the brain that has a lot of mucus and very little blood, since the medium would be too dense and sluggish: "For mechanics teaches that in a clock the wheels directly put in motion by the spirits are the finest and quickest of all." Therefore, the motive principle of the human body must reside in a volatile, quick medium, such as the nerves. "The stimuli of all perturbations or feelings of the soul are the appetites of desire and anger; blood seems to be the vehicle of desire, bile of anger."[98] The intermediary between mind, which resides in the spirit, and spirit, which resides in the passions, is the "mind of spirit," the *mens animi.*

A passage from the autobiography serves to ground the physics of the *mens animi* into a coherent metaphysics. After reading Bacon's treatise *On the Wisdom of the Ancients* ("more ingenious than true") Vico was inspired to attempt an etymological reconstruction of the ancient philosophy of the Italians, going back earlier than Pythagoras, even, to his Egyptian source. To follow his researches we must start with the word *coelum*, which for the Latins means both "chisel" and the "great body of the air." Imagine that the instrument with which nature makes everything is the wedge, and that this is what the Egyptians meant their pyramids to signify. The Latins called "nature" *ingenium*, whose property is "acutezza" or sharpness, as in a "sharp wit." "Nature forms and deforms every form with the chisel of air. It would form matter by shaving lightly, and deform it by gouging deeply, with the chisel by which the air ravages everything. The hand that moves this instrument would be the ether, whose mind by all accounts was Jove."[99]

Air is the passive, female principle that gives the universe motion and life; ether is the male principle that acts on it. Ether is insinuated into the nerves and is the principle of sensation in man, otherwise called the animus. This is what moderns call animal spirits. Air is insinuated into the blood and is the principle of life in man, otherwise called the anima. The moderns call it vital spirits. "In proportion as ether is more active than air, the animal spirits would be more mobile and quick than the vital." Now, just as male ether acts on female air, mind works on ether. This is the *mens* of the animus, the spirit-mind that comes to us "from Jove, who is the mind of the ether." What begins as an etymological inquiry results in a metaphysics based on an original physiology.

Vico's familiarity with the latest medical theories, the fact that he had well-developed, informed opinions regarding physiology and disease does not imply that he was entirely a modernist. A section in *Il Metodo degli studi del tempo nostro* is dedicated to medical studies. In it he deplores the lack of energy devoted to pathology or semiotics. Just because the art of healing is an uncertain science does not imply that we should give up on understanding and pinpointing the causes of illnesses. Only investigative diagnosis, founded upon long observation, can achieve "a sure curative system." And he adds that even if "the Galenists did not succeed with their syllogizing at conjecturing with exactitude the causes of illnesses, likewise I would say that neither do the modern physicians with their soritizing hit the mark."[100]

This is not to suggest that Vico blindly followed the ancient theory of the humors. He did seem to take for granted the kind of macro-microcosmic correspondences familiar to students of physiognomy that see physical symptoms as indicative of moral character: in a sure, curative system, he says, there is a correlation between symptom and cause that is "in every bit as precise as the correlation between illnesses of the body and those of the soul."[101] But then, so do we postmoderns. We judge fat people to be lacking in self-control, for instance. We judge women to be changeable and moody, especially around the time of their period. These kinds of assumptions formed part of his late Renaissance culture as they do ours.

In Nicola Solla's brief life of Vico, the disciple describes the physical characteristics of his former teacher and friend: "He was of medium height, of an adust habit of body, his nose aquiline, his eyes lively and penetrating, and from their fire it was easy to conceive the power and energy of his vigorous mind. His choleric temperament contributed to the sublimity and swiftness of his intellect."[102] The science of physiognomy, discussed in the last chapter, gives us the necessary interpretive keys to decode this portrait. Our informant for this physiognomic diagnosis will be Francesco Stelluti.

Vico's height, neither too tall nor too short, would indicate overall harmony of bodily proportions as befitting a just man. *Mediocrità* is the sign of men of prudence, liberality, magnanimity, wisdom, and especially quick-witted genius.[103] In addition to "a wholly well-proportioned body" the just are said to have dark hair, as Vico did. Solla writes that Vico's eyes were lively, penetrating, and fiery, associating their brilliance with the power and energy of his vigorous mind. Stelluti describes the just man as having "big eyes, high, prominent, shining . . . with the inner circle around the pupil narrow and black and the outer circle fiery." Vico's aquiline nose comes under the rubric of the eagle, evidenced in "a beaked nose, prominently protruding from the forehead." The behavior of aquiline men was believed to be "fierce, warriorlike, magnanimous, generous, even-tempered, predatory, & spirited."[104] Solla confirms this diagnosis, stressing Vico's quickness to respond to his attackers, painting the portrait of a warrior soul tempered by greatness of spirit, tenderness, and generosity toward his friends: "He loved his intimates with an excess of tenderness, preferring a respectful friendship to servile fear. He was discreet and indulgent in censure, sincere and liberal in praise. If attacked, he defended himself, but always within the bounds of decency and equity."[105]

Vico's choleric temperament was evident to all who knew him; it plagued the philosopher himself, who fought all his life to control his irascible temper. A bitter, melancholic disposition was the cause for "the sublimity and swiftness of his intellect," as Vico himself points out in his autobiography.[106] A choleric temperament, dominated by yellow bile, mixed with melancholic humor or black bile, is certainly what stands behind his "adust habit of body."[107] Melancholic humor was associated with warm and subtle blood, indicative of "grande ingegno [great genius]." When the blood is also delicate, as Vico's surely was, it is further proof of intelligence.[108]

Stelluti informs us that melancholic humor is manifested in thinness and prominent eyes and an avaricious, astute, timid, seditious, and envious character. Vico might have been overly melancholic in his youth, starting from the fateful fall and evidenced in his gloomy moods and slenderness. But although still plagued by excessive yellow bile, by the time he reached adulthood he had clearly tempered his melancholic black bile. By cooling his black bile Vico achieved exactly what Stelluti would have predicted: when the heat is cooled down it leads to a temperament that is "extremely prudent, surpassing others in humanistic studies, in the military arts, and in government." When the melancholic humor is tempered as it was for Vico "the man will be of excellent judgment, & will have an extraordinary intellect."[109] We would have to conclude that the fourth, phlegmatic humor—associated with humidity and coldness, sleepiness, and laziness in studies—held little sway in Vico's temperament.

Solla's biography also provides a physical picture of Vico's later years and gives a sense of the aging man, who for several decades had been known by the nickname "Master Tizzicussus."[110] Solla describes Vico's last days: "Already some time before his death the philosopher had completely lost his memory, since wanting to reason about some affair or another, he would suddenly become seized, mute, and stupid." When "shaken out of his stupor by a servant, who would act as his interpreter, he would indicate by affirming or negating with gestures and with his voice as if he had fished back out of oblivion an idea previously submerged and lost."[111] How ironic that the man of ingenuity and depth "quick as lightning in perception"[112] should end his days with the symptoms similar to what we call Alzheimer's disease!

In his own estimation, by 1740, four years before his death, the torturous war between body and soul first documented in his "Affetti" poem had taken its irreparable toll. Vico is writing to Carlo di Borbone to grant the chair of rhetoric to his son Gennaro, who had already replaced his father as lecturer some years before. The letter begins with a frequently heard lament regarding the pitifully small salary, which had never been sufficient to maintain Vico's large family. He goes on to describe in third person his physical state: "And because he has now reached a very advanced age [72 years old] and is aggravated and almost oppressed by all those evils [*mali*] that the years and continual toils that he has suffered are wont to impart . . . such as illnesses [*mali*] of the body, accompanied and united with more powerful ones, such as are those of the soul, they have left him in a truly incapacitated state for life, no longer being able to drag his body about, by now tired and almost ruined [*cadente*], so that he lives miserably, practically nailed to a bed."[113] The recurrence of the lexicon of his youthful poem, the repetition of *mali*, evils and illnesses of body and soul, and the description of his body as *cadente*, ruined, aged, shows that Vico's struggle with physical pain had been a continuous theme throughout his life.

In another passage at the end of his *Autobiography* he gives an idea of the illnesses that wracked his body: "Because of his advanced age, worn out by so many labors, afflicted by so many domestic concerns and tormented by spasmatic pains in his thighs and legs and by a peculiar illness that devoured almost all that lies between the lower bone of the head and the palate, he completely gave up his studies."[114]

In spite of these references to poor health, Nicolini refuses to give credence to the exaggerated tales of Vico's last days, told by Villarosa and added to the English edition of the *Autobiography*. Nicolini does not include the Villarosa continuation of Vico's last years written in 1818, because it is full of errors, but also, I think, because it is repugnant to a lover of Vico to imagine the great thinker's last days of dotage. Mediating between Solla's southern Italian propensity to exaggerate and the fact of Vico's continued literary output during the last fourteen months of his life, Nicolini is willing to accept only

> that Vico, afflicted throughout his entire life by all sorts of ailments, perhaps remained after the death of his eldest son Ignazio, weary in body and spirit; that perhaps he precipitated ever more toward misanthropy and taciturnity; that, along with old age, what were always the negative qualities of his mind—confusion, memory lapses, "illnesses of inexactitude," lack of diligence, poor acuity, and relative slowness—became ever more visible. . . . But that that supreme intellect could ever entirely be extinguished, that that giant of thought could ever become an inert torso, is even less admissible considering, as we have said, that his manuscripts prove the contrary to be true.[115]

A giant of thought become an inert torso, what an unnatural progression from mobility to stasis!

I think it probable that Vico suffered from advanced Hypochondria, a malady of the stomach. In an English treatise written in 1711 on the *Hypochondriack and Hysterical Diseases*, Bernard Mandeville describes the major interlocutor in his dialogue, Misomedon. The description fits neatly with Vico's character and symptoms:

> Misomedon is a Man of Learning; who, whilst he had his Health, was of a gay, even Temper, and a friendly open Disposition; but having long labour'd under the *Hypochondriack* Passion is now much alter'd for the worse; and become peevish, fickle, censorious, and mistrustful. Notwithstanding this, in fine Weather sometimes he has lucid

Intervals; that last for two or three Days: In these he is very talkative, loves to converse with Men of Letters; and is often facetious. Tho' in his Discourse, he seems not to want Sense or Penetration; yet he is partial in his Censures, and unsteady in his Humour; sometimes very complaisant, at others captious; but always prone to Satyr. He is much given to ramble from one thing to another; and often to change his Opinion. When he speaks of himself, he is apt to break out in Rhetorical Flights; and seems to take Pleasure of talking of his Ailments, and relating the History of his Distemper; and what has happen'd to him. He has study'd Physick, but is no Well-wisher to it, and bears a great Hatred to Apothecaries.[116]

Another symptom of Hypochondria, "when the Distemper has been of a long standing" is rather "odd and singular," and definitely one that the pedantic Vico suffered from: an over fondness for quoting Latin Proverbs makes the Hypochondriac "fuller in his Discourse of Quotations from the Classicks, than a Man of Sense, that understands the World, would chuse to be, if his Head was perfectly clear."[117]

The character who suffers from advanced Hypochondria relates the terror of never knowing when his mind will snap:

'Tis Heaven to me when I think how perfectly well I am; but then how miserable on the other side again is the Thought, of harbouring some where within me, tho' now I feel it not, a vast, enormous Monster, whose Savage force may in an Instant bear down my Reason, Judgment, and all their boasted Strength before it. . . . I know it, I resist it, yet I can't overcome it; and when it begins to be violent, I must apply to my self in sad earnest, what Phaedria in the Eunuchus said on a foolish Occasion: *I see my Ruin before-hand, and I can't help knowingly to run into it.*[118]

His inability to resist the ruin that he sees, knows, yet succumbs to, reads like a recipe for Vico's fool: without control over fear and desire, the mind creates chimeras, monsters that bear down on reason and judgment. Misomedon is such a fool, allowing the disease of nerves and organs of the body to pervert his mind. The result of submitting to the body is the birth of an enemy to reason, a savage monster of violent powers that corrupts one's free will.

It seems likely to me that Vico suffered from the same advanced Hypochondria that also killed Tommaso Cornelio. It was a disease typical of literary men of genius, especially those with melancholy temperaments. Mandeville adds that these patients "are oftner Men of Learning, than not; insomuch, that the *Passio Hypochondriaca* in High Dutch is call'd . . .

the Disease of the Learned." He disagrees with his English colleague Thomas Willis regarding the role of the spleen in this disease. The reason why men of sense, especially those of learning, are more subject to Hypochondria is that they "are guilty of Errors, that, unless they are of a very happy Constitution, will infallibly bring the Disease upon them, such as Blockheads can't commit; for [they are] all Men that continually fatigue their Heads with intense Thought and Study, whilst they neglect to give the other Parts of their Bodies the Exercise they require."[119]

That Vico would subscribe to my diagnosis is probable, I think, from the example of two of his own diagnoses. One is brief and regards Father Don Roberto Sostegni, Florentine canon of the Lateran: "Though suffering from an excess of the choleric humor (which often made him mortally ill and finally, abscessing in his right side, caused his death . . .), he nevertheless controlled his temperament so well by his wisdom that he seemed by nature the mildest of men."[120] The second diagnosis regards himself:

> Vico was guilty of being choleric [*pecco' nella collera*], from which he guarded himself as best he could in his writing; and he publicly confessed to being defective in this: that he would inveigh too passionately against the errors of thought or of doctrine or against the misconduct of those men of letters who were his rivals, whilst as a true philosopher with Christian charity he should rather have dissimulated or pitied them. . . . Among the semi-learned or pseudo-learned, both because they were poor scholars, the more shameless used to call him crazy, or in somewhat more civil terms they would say he was eccentric and had odd or obscure ideas.[121]

Both of Vico's diagnoses stress the inevitability of even the most magnanimous soul falling or lapsing into weakness through "a disease of the nerves, a defect of the organs, or an intemperate desire," the phrase from the Second Oration spoken in the context of the fool's struggle for internal harmony. For Vico the most heroic and praiseworthy struggle, the only battle that a Christian could fight, was that of dominating the bodily passions through wisdom. The fool or *stultus* becomes wise or *sapiens* by finding a state of natural justice between body and soul. We are all corrupt, we are all subject to desires and diseases. Only men or women of wisdom will dominate their bestial tendencies, bringing their bodies and minds into line with the beautiful light of spirit.

This is the task he gave himself after the bout of severe melancholy suffered in his twenties and passionately documented in the "Affetti" poem. The fruit of that suffering was a theory of natural justice that explained the motions of the body and the motions of the body politic in terms of progressive adequation to the precepts of reason as revealed in the working of divine providence through the agency of *conatus*.

Vico's Medicine: Virile Choler and
the Death of Angiola Cimini

The most revealing source for Vico's physiology does not come from his philosophical or historical writings. To bring to light the beliefs he took for granted to the extent that they did not require any explicit mention or theoretical metacommentary (Foucault's *episteme*), we must turn to a funeral eulogy that he wrote for Angiola Cimini, Marchesa della Petrella, in 1727, two years after completing the first version of the *Scienza nuova*. In this little gem of eloquence, judged by one misguided admirer to be superior to the *New Science*,[122] Vico develops a "profound speculative" theme, one of the fundamental motifs of his thought, according to Benedetto Croce: the revindication of the positive value of the passions.[123] We could read this oration as a philosophical speculation on "heroic fury," as Croce does; or we could admire it as a powerful piece of rhetorical eloquence; or we could appreciate it as a heartfelt expression of grief for a young woman he loved as a daughter. We will be reading it as a source for Vico's physiological and medical thought.[124]

Angiola Cimini and Giambattista Vico had much in common. They were both endowed with acute intellects. They both loved poetry, philosophy, and history as vehicles to wisdom. They both suffered from delicate complexions, fragile health, and atrocious maladies of body and soul. They were united in friendship through Vico's daughter, Luisa, and they conversed frequently in the company of other female poets, literary friends, and intellectuals who gathered around "this impulsive, delicate lady of genius," meeting mostly in her salon or at times in Vico's own house.[125] But the main thing they shared was an irascible temperament and a similar approach to mastering it.

Like Vico, Angiola Cimini suffered from choler. In her case, the yellow bile was so violent that as a small child if she failed to obtain her slightest desire, she became so vexed and worked up that she would "throw herself upon the ground, afflicted in all her parts, to the point of banging her tender head on the hard floor."[126] Vico explains in compassionate detail the physiological causes for her defective character:

> When what physicians call "lymph" abounds, being insipid and lazy, it makes men patient and phlegmatic; when the vital blood is too exuberant and effulgent, lightheartedness—knowing only how to depict beautiful hopes and inducements—makes not only objects of

hazardous outcome but even the most grievous and dire events appear figured with joyful countenances; when that slow and sticky juice that makes men melancholy (leaving it to the doctors to dispute the term) is excessive, it makes the souls of these men reserved and weighty in the face of life's adversities. But superabundant choler—being the sulphurous part of the blood—at times kindles a fiercely burning fire of spirited bodies; since the sicknesses that it causes are all acute, precipitous, fatal, similarly, the perturbations which agitate the soul are unbridled, blind, extremely violent: . . . thus, in order to cure an acute choleric passion superhuman virtue is required, what the Greek poets with great knowledge of the senses and equally dignified expression called "heroic."[127]

From this passage we may say with confidence that Vico believed implicitly in the theory of humors and that he understood their functioning in physical, scientific terms. He does not present himself as a medical expert and has no interest in the latest debates on the definition of melancholy, but what he does believe without any doubt is that the passions arise from the movements and degree of agitation and temperature of the animate particles that course through blood, bile, and phlegm.

He runs through the four humors in this order: phlegm, blood, black bile, yellow bile. For each humor he provides a material description and then connects the physics of its movements with the passions of the soul. When phlegm is slowed by lymph it makes a man patient. When the blood is overly luminous, the sanguine mind perceives only the light-hearted side of life. When the black bile becomes adhesive and thick, men become melancholy; their hearts are heavy and hesitant in the face of adversity. When yellow bile dominates the humors its sulphurous nature catches fire, causing violent perturbations of the corpuscles, and acute distress to the soul as evidenced in fatal, painful diseases symptomatic of choler. Even though he understands the humors as chemical substances, in the same way his spagyrist friend Donzelli would, the associations he makes with corresponding passions are analogically related as a physiognomist's would be. He syllogizes in the manner of the Galenists rather than soritizing in the manner of the moderns, we might say.

Luminosity of the blood can be connected both with Descartes's "ordinary fire " that heats up the heart as well as Van Helmont's life principle that he called *Archeus,* and which Sebastiano Bartoli wrote a book about.[128] The identification of yellow bile with sulphur is reminiscent of Borelli's opinion that animal spirits were material substances combusted in the blood, similar to what happens when oil of tartar is poured onto spirit of vitriol. The association between fiery heat and violent maladies, such as fevers, is repeated in Vico's debut lecture to the Academia Medinceli in 1698.

The context is a pretty piece of philological erudition on the "Sumptuous Dinners of the Romans."[129] In an attempt to explain how the Romans managed to preserve wines for a hundred years or more, he attributes their success to the smoke that they fanned around the tar-covered storing vases. He proves his assertion by the practice of wearing pitch clothing to protect oneself from the plague. The pitch blocks malignant air from introducing itself into the veins, where it causes fever. And fever, he says, citing Thomas Willis, the English physician who discovered the same functioning of the digestive system that Tommaso Cornelio did, is nothing but a fermentation of the blood, resembling fermentation of wine.[130]

Just as smoky flames cure wine through fermentation, sulphurous spirits provoke fever in the blood. This is also Borelli's opinion regarding the origin of diseases: a small amount of ferment in the blood that is difficult to remove. Another echo of Borelli's theories on the *succo nerveo* is evidenced in Vico's use of the word *sugo* to describe the melancholic humor. Vico's medicine appears to be a potpourri of the most modern and most ancient sources.

Having ascertained the cause of her illness, Vico goes on to describe what Angiola did to cure herself from the savage lion that was feasting on her insides. Starting from the time she was a small child, she began to combat that "raging savage enemy and to tame it by the use of virtue." She began to refuse all food. When placed at the table by her nurse, Angiola would become sad, then burst into tears, "ready to die of hunger rather than take the smallest taste of food."[131]

Vico explains her behavior with another detailed medical explanation. In her stomach, says Vico, Angiola had

extraordinary ferocity or resentment, because, since her temperament was, unfortunately, faulty in its choler, that part in our viscera where choler acquires its principal functions must have been in her of the same wild and arrogant nature—which is why the authors of Greek speech, which was the language of philosophers, called the "stomach," with heroic and almost natural expression, the hypochondria—so, being sick with such a disease, and being unable to come to terms with anyone in her household, nor to taste any foods, no matter how exquisite or sumptuous . . . which later caused terrible illnesses and eventually her death, she began to practice severe penitences in order to break the arrogance of this savage lion that feasts inside the chest of choleric persons.[132]

Speaking as a choleric himself, Vico must have meditated long and hard during his bouts of Hypochondria on the physical and spiritual causes for his illness. This is the explanation that he evidently found most plausible. As he said in his section on medical studies, there is a precise correlation between the diseases of the body and those of the soul. Similarly, there is a natural relation between the stomach and Hypochondria, so much so that the Greeks had one word for both. By starving the voracious lion of pride in her stomach, Angiola was correctly responding to the signs of her fallen nature manifested physically in her viscera. The process of domesticating the beast within must take place in both the moral and spiritual realms. Accordingly she refused to eat and began to practice severe penitences. This Vico sees to be the correct response to her faulty nature because piety and religion are the base and foundation of all moral and civil virtues.

The other therapeutic approach she adopted involved the study of letters. Eventually, by following this twofold program, the practice of piety and the development of the imaginative and reasoning faculties, she transformed her savage nature into a docile, truly human semblance with such a degree of success that those who met her would have believed her to be phlegmatic rather than choleric:

> And more and more she weakened and conquered it with the study of letters, and especially with the exercise of Christian piety, with which she domesticated it to such a degree that, grown into womanhood, anyone who had never met her before, unless he were an expert philosopher in the signs of human comportment—which in her from the agile, quick movements of her body and her rapid, swift comportment, would have indicated that she had a certain spirit and fire that pointed to her true nature—seeing her seated, from her relaxed, modest deeds, from the peaceful turns of her ever serene eyes, from her most delightful, never strident or hurried words, and from her senses, altered into humility and most noble meekness, he would certainly have believed her to be phlegmatic rather than not.[133]

The parallel between Angiola's history of humanization and the history of human origins described in the *New Science* is an exact one: the monster that resides in the physical body, and shapes it accordingly, responds to the light of providence and is dominated by penitences, ritual worship, ablutions, and the development of reason. This is another way of describing how divine providence brings gentiles into a state of natural justice. This passage also speaks of Vico's faith in the discernment of physiognomists, whom he dubs "philosophers of the signs of human comportment." He proceeds to read Angiola's body and movements according to physiognomic precepts and reiterates a traditional interpretation according to the humors.

I have no doubt that Vico was a lover of beautiful women. He loved beauty as much as he loved virtue and wisdom. A large portion of Angiola's funeral oration is devoted to her "corporali doti [physical endowments]," offered as proof of her superior intellect and virtue. Her body displayed, by all accounts, an unearthly beauty. Vico expounds at length about her "beautiful and delicate body," her "graceful and delicate members," perfectly proportioned and fitted together in sublime harmony, her "delicate and gentle complexion, tinted with a vermilion hue that appeared and disappeared, which is the sweetness of the color that Aristotle defines as the fulfillment of beauty." Angiola was further endowed with a quick, agile movement that lent a vivacious grace to her face and a dignified demeanor to her posture. Nature, responding to her virtue, also granted her a most sweetly pleasing voice "that indicated the well-regulated measures of the very beautiful body from which it issued."[134]

However, as an inevitable sign of the limits to perfection that characterize "our common, unhappy human condition" her defective nature soon began to show. Angiola's battle with excessive choler is what led her to exercise superhuman acts of will, making of herself "a woman of heroic virtue." Her defect, violent attacks of rage, provided the material upon which later she could develop and exercise her sublime, youthful virtue. This paradox founds the logic of humanization in the *New Science:* our unhappy human condition is the raw, defective material upon which the will, through the practice of virtue, is encouraged to assert itself.

Fear and ignorance, not detachment and knowledge, are what led man to make himself into his true, human form. The birth of divinities came about not through philosophical reasoning but through fear, "not fear awakened in men by other men, but fear awakened in men by themselves." The birth of divination came with gruesome, sacrificial rites: "For all this was necessary to tame the sons of the cyclopes and reduce them to the humanity of an Aristides, a Socrates, a Laelius, and a Scipio Africanus."[135] What is reduced in the sons of the cyclopes is both their deformed, gigantic bodies and their raging, violent spirits.

Man becomes all things, not by knowing them but by *not* understanding them: "For when man understands he extends his mind and takes in the things, but when he does not understand he makes the things out of himself and becomes them by transforming himself into them."[136] The story that Vico tells of Angiola Cimini is about how she transformed herself into something that she didn't understand at first, but only came to know while she did it. The defect with which nature marred her beauty was at the same time her instrument for developing herself into a superior human being.

Vico describes her physical presence as a healing one for those who came into contact with her. Sent to be educated in a convent school for noble Spanish ladies, Angiola was adored by her classmates, who came to her to be healed of their ailments. He says that she did this using the ancient techniques of medicine that only Pythagoras truly understood: "With the pleasing harmony, not of song and sound, but with the kind that Pythagoras alone in the world understood, that issues from a most vivacious spirit delicately tuned with a beautiful and graceful body, from this she tamed the ferocity of diseases, soothed their torments, relieved their pains."[137] She did this simply through her conversation, thus demonstrating that although man is fallen in his mind, his spirit, and his speech, the healing power that emanates through speech from the perfectly tuned harmony of body and spirit is proof of our ability to pick ourselves up from those falls, to make of ourselves what we imagine we could be.

Unfortunately, Angiola also erred on the side of excessive virility, ultimately causing her death at the age of twenty-seven. Already when she was young, Vico says, you could tell that her attacks of rage were "reasonable and generous as befitting a woman of heroic virtue" and not "reckless and at times even savage, common to all the female sex." A combination of excessive "virile choler" and a superior intellect, unsuited to the female body, caused Angiola to repeatedly miscarry late in her pregnancies. After the first miscarriage she miraculously recovered from malign fevers but was subsequently drawn into "bad habits from maladies of the body": a spasmatic gushing hemorrhage made her so weak that for a long time she was unable to hold down any food. In order to survive she began to crave unhealthy foods, which generated "sughi viziosi," corrupt juices of the humors. The third pregnancy was her last; she died eight days after the miscarriage. The medical explanation Vico offers for this series of events is as follows:

> So that because of a certain natural course of things, a fact often confirmed by experience, women furnished with a much greater intelligence than [normal for] the female sex are less suited to procreate, perhaps because this requires of them that their body fabric be of an extremely loose weave, so that their viscera can yield to the growing fetuses, and even more in order to give birth to them, which is the task given to women by providence; that is why when it comes to the enjoyment of the senses women are marvelously judicious, they have robust powers of imagination, and they are much more

fussy and extravagant in taste than men when it comes to pleasures and delicacies. Because sensible objects sink deeply into the folds of their brains, like in liquid wax, they feel [or perceive] very distinctly whatever is pleasing or repugnant to them; the contrary is true for the exercise of strength, a manly virtue, calling for a contrary force of reflection, that holds the animal spirits tight against the pleasures of life, and inures them to toil, pain, and death: because of all this, if I do not err, virile choler, in which she abounded, snatching away the humidity that is necessary to nourish fetuses when they become large, brought to pass that, as ill luck would have it, in the sixth month, a fatal one according to doctors, she would miscarry every one of them.[138]

This is the only example we have of a fully developed diagnosis by Vico according to his new system of medicine based on the "distinctive me-chanical medicine" of the Egyptians, that of "slack and tight." Vico's modern adaptation is based on Descartes's definitions of heat and cold: "cold being motion inward from without and heat the reverse motion outward from within."[139] The heart is the heat-producing furnace at the center of the animate body. When the heart lacks air, its motion becomes weaker; the slow-moving blood begins to clot, "which is the chief cause of acute fevers."[140] In Angiola's case, excessive heat caused the body weave to tighten, whereas in a woman it should have remained slack. Because she nurtured her intellect instead of her womb, her body would never be moist or slack enough to bring a fully matured baby into the world. She died a victim to the principles of slack and tight.

Vico ends the story of her last days with an apt remark: "We do not depict Angiola Cimini Marchesana della Petrella as we imagine her, but we discuss her as we understand her."[141] In other words, this portrait is the fruit of reason, not fantasy. It seeks, like his autobiography, to chart the workings of divine providence in the ontogeny of one singular woman. The phylogeny of the human species echoes the ontogeny of Angiola Cimini.

Conatus

"Nature is motion. The indefinite power to move underlying this motion is *conatus*. The infinite, and in itself motionless, mind that excites this power is God. The works of nature are brought into being by *conatus* and brought to perfection by motion. In sum, the genesis of things pre-supposes motion, motion presupposes *conatus*, and *conatus* presupposes God." This is how Vico sums up his argument in his *Liber metaphysicus*, after demonstrating that *conatus* is a shoddy concept for physics, no better than "natural sympathies and antipathies" or "hidden qualities." "So in order for the language of physics to be perfected, *conatus* should be taken out of the schools of physics and restored to the metaphysicians."[142] *Conatus*, in a nutshell, is not motion, it is the metaphysical principle of motion, just as a point is not extension, but rather the indefinite power of extension.

Between the immaterial and the material substances lie the air, the ether, the nerves, and the blood:

> Moreover, the vehicle of sensations is the air that insinuates itself into the nerves and stirs their sap, and distends, inflates, and twists the fibers. Nowadays in the Schools the air that moves the blood in the heart and arteries is called *spiritus vitales*, and that which moves the nerves and their sap and filaments is called *spiritus animales*. But the motion of the animal spirit is far quicker than that of the vital spirit, for as soon as you want, you can move a finger, whereas the blood takes a long time—some physicists calculate that it takes a third of an hour to pass from the heart to the finger in its circulation. What is more, the heart muscles are contracted and dilated by nerves, so that through systole and diastole the blood is made to circulate continuously, receiving from the nerves its own motion. Therefore, they called this masculine and strenuous motion of air through the nerves the animus; but the feminine submissive motion of air in the blood which overlies it, so to speak, they called the anima.[143]

As we have seen in detail from Vico's medicine, the will is a movement of the *mens animi*, which also moves the passions. Since the spirit resides in the passions, by an act of will—as Angiola Cimini demonstrated through her "sweet austerity"—we can make ourselves what we imagine ourselves to be: "I seem to myself to be the same person, but from one moment to the next I am a different person because of the constant coming and going of the things that enter and leave me," says Vico. Because all bodies are in constant motion, and all motions are produced by air pressure from all sides, there can be no such thing as perfectly straight motion, for instance. "Air is the vehicle of life, which, by its inhalation and exhalation, moves the heart and arteries, and the blood in them. This movement of the blood is life itself."[144]

If we begin with the assumption that air is the pressure of "God's perceptible hand" and we add to it another, that "all things are moved by perpetual motion and that there is no rest in nature," it is easier to understand Vico's assertion that "it is not within our power to move anything. God alone originates all motion and arouses *conatus*, which is the beginning of motion. It is the determination of motion that is truly within our power." We cannot make anything really, we can only put things together by determining their motion. We can, however, and unlike the brutes, discern "what is apt, fitting, beautiful, and ugly." We are the gods of artifacts: "Nature generates physical things, so human wit gives birth to mechanics and, as God is nature's artificer, so man is the god of artifacts."[145]

Our examination of monstrosity has made a full turn. The monster is that which should be passive due to its inert nature, yet inexplicably moves as if it were self-moving. The automaton is an example of something that moves as if it were animate and yet is a machine. Had Vico entirely accepted modern mechanistic medicine he would have been forced to conclude that our bodies are nothing but faulty machines, and the soul nothing but material movements of animal spirits. This would be tantamount to making man a monster, a brute, and his technological creations so many other monstrosities.

But through the influence of Neapolitan scientists and especially Paolo Matia Doria, by transferring the principle of motion out of the material realm and into the immaterial, Vico was able to establish a chain of command that ran equally, in parallel fashion, between God and the human body in the microcosmic realm, and between Divine Providence and the civil body of nations in the macrocosmic realm of history. There is a spark of divinity in all beings of the created world, including the creations of mankind. The chain of command outlines the principle of motion in both the individual body and the larger world of history. It runs like this:

> You have here the doctrine that the spirit rules man's soul, his mind rules his spirit, and that God rules the mind. The mind makes fictions by paying attention, or the human mind makes truths *ex hypothesi,* while the divine mind makes true things absolutely. Hence, man is given his mother wit for knowing or making. Finally, you have the doctrine that God wills by a nod or by making. He makes by speaking or by the eternal order of causes, which we, in our ignorance, call chance, and in our self-interest, we call it fortune.[146]

The crucial question for both the physical and metaphysical realms concerns the problematic nexus between matter and consciousness, motion and will. If we were inert, passively moved brutes there would be no hope for redemption. But we are not. We are the gods of artifacts, and that includes our bodies and our selves. Man is given his mother wit for knowing or making, by means of which we make ourselves. We *are* self-moving. There is no possibility of confusing ourselves with animals or machines so long as we grasp the way in which God's invisible hand becomes perceptible through the air and through the ether. "Even in our errors, we cannot lose sight of God. For we embrace falseness under the guise of truth and evil under the likeness of goodness. . . . We seem to see motions started and communicated by bodies, but this very stirring of motion, these very communications, assert and prove God—God as mind and as author of motion."[147]

We humans are moved in mysterious ways through God's will in the same way fire, plants, and animals are moved: the motion common to air becomes the motion proper to each creature's "particular mechanism." The way that God's will comes to our spirit is through our mind, "which is the particular form of each of us. So that every act of our will is both our own true and proper choice and, at the same time, the infallible decree of God."[148] This physiology is integrated not only into a metaphysics but also into a theory of human history. Monstrosity in Vico is the tale of human becoming, out of body into consciousness, out of the sensible into the divine light of reason. It is the founding tale of our human origins.

The Physician of History: Monstrosity in the New Science

In his *New Science* Vico offers the following advice on how the body politic should be understood:

> The academies . . . should teach the young that it is the nature of the civil world, which is the world that has been made by men, to have just such matter and form as men themselves have. And, accordingly, that of the two principles that compose the civil world, the one is of the same nature and has the same properties as our body, the other as our rational soul has—of which two parts the first is the matter and the second the form of man.
>
> It is by being prepared in this way that the youths to be taught will learn the practic of this Science founded on the eternal law that providence has established for the world of nations. This is that nations are secure and flourish in felicity so long as the body in them serves and the mind commands.[149]

This passage would read as an extended metaphor or an apt conceit were we not familiar with the physics that underlies Vico's comparison between the body and mind of man and those of nations. The material part of the body politic is made of the foolish *stolti*: "Hence the men who have neither counsel nor virtue of their own are the matter which is the body of the world of nations." These are the formless men, allied with defectiveness or vice: they are sluggish and inert, lazy and effeminate: "all the delicate soft and dissolute idlers." In a word, they are the fools, the doltish, "who are always changing their minds, who never remain content with the same thing, who always love and affect new things . . . whose fixed epithet is *mobile.*"[150] They are mobile, not because they are a "mob," as the translators suggest, but because they are brutes, animals moved by outside forces, perpetually buffeted by the winds of their desires. They are those who have given up their human will, their autonomous self-moving principle of motion. Consequently, the fools make up the passive, material part of the body of nations, drawing the society of men toward disorder and chaos.

The chaos of the theological poets has already been shown, says Vico, "to signify the confusion of human seed." Confusion of human seed signifies the "nefarious bestial life when this earth was an infamous jungle of beasts." The fool is thus further revealed as the distant descendant of the grotesquely proportioned children of Noah, who "having lapsed into a state of bestiality, went wandering like wild beasts until they were scattered and dispersed through the great forest of the earth . . . when the heavens thundered for the first time after the flood." "The giants were by nature of enormous build, like those gross wild creatures which travelers report finding at the foot of America, in the country of the so-called Patagones [Big Foot]."[151]

In our view, Vico says, they could only have arrived at their monstrous statures through the bestial education of their children. He later describes what those educational practices were:

> By fleeing from the wild beasts with which the great forest must have abounded, and by pursuing women, who in that state must have been wild, indocile, and shy, they became separated from each other in their search for food and water. Mothers abandoned their children, who in time must have come to grow up without ever hearing a human voice, much less learning any human custom, and thus descended to a state truly bestial and savage. Mothers, like beasts, must merely have nursed their babies, let them wallow naked in their own filth, and abandoned them for good as soon as they were weaned. And these children, who had to wallow in their own filth, whose nitrous salts richly fertilized the fields, and who had to exert themselves to penetrate the great forest, grown extremely dense from the flood, would flex and contract their muscles in these exertions, and thus absorb nitrous salts into their bodies in greater abundance. . . . They must therefore have grown robust, vigorous, excessively big in brawn and bone, to the point of becoming giants.[152]

Nitrous salts, otherwise known as sulphur, were the material cause for the gigantic growth of the feral babies. As we have seen, many authors of physiology posited that the life force resided in a fiery substance that ran through the blood and nerves. Many believed that the fire was fueled by a nutritive element, niter or sulphur.[153] Vico himself identified the choleric humor as a sulphurous component of the blood. If a baby absorbed through its nerve endings leading to the heart and eventually to the brain an abundance of niter coming by way of its own excrement, it would grow robust in both body and brain, feeding its spirits with pure combustive power and developing a vigorous, active imagination. The moral cause for the giants' grotesque proportions lay in their isolation, their lack of contact with human customs and human speech.

The immaterial part of the body politic, its reasoning soul, is made up of the men of form, "who can counsel and defend themselves and others, and these are the wise and strong."[154] The *sapiens* are the luminous beings, who are industrious and diligent, men of propriety and dignity, such as the scholar, the jurisconsult, the artisan. Vico stresses their seriousness, their gravity, their constancy. These are the self-movers, whose spirits move with the rhythms of the celestial bodies. They are those who have the property of being "lo stesso": both themselves and the same.

The regularity and ordered nature of their movements leads to beauty and harmony, and in an ordered state unity flourishes. Vico makes clear that all men are weak and corrupt and that their natures alone would not be sufficient to bring them to such civic virtue. He refers to the humanization process as an "exertion" or "sforzo" that only some men or women (like Angiola Cimini) are capable of. Religion and laws, assisted by the force of arms, may move some men to found humanity, and it was only a few of the more robust giants who did.

The question we are asking, then, is not what distinguishes a man from a monster. The answer to that question lies in the principle of movement that animates one or the other. What Vico asks next is: what is the motive force in the individual and in the race that educes the truly human out of the monstrous savage? The answer to this question also lies in the principle of motion, applied now as both a physical and a metaphysical force to the body and mind.

"Moral virtue began, as it must, from *conatus*."[155] *Conatus* is defined in this context as "control over the motion of their bodies." It imposed "form and measure on the bestial passions of these lost men and thus transformed them into human passions." The first locus of transformation thus resides in the motion of the passions, which reside, as we have seen, in the spirit. The agent of control is the mind of the spirit, the *mens animi*. "This control over the motion of their bodies is certainly an effect of the freedom of human choice, and thus of free will, which is the home and seat of all the virtues, and among the others of justice."[156]

In other words, following the same paradox that we identified earlier in relation to Angiola Cimini, gentile mankind began in monstrous form in order that, through the exercise of virtue, he would discover his free will and remake his human form according to the lineaments of natural justice. "Man alone is whatever he chooses to be. He becomes whatever he desires to be."[157] To be human means to understand that you are your own master, that you have the choice to make yourself in your own image. The bestial and the foolish are akin to machines precisely because they have no motive force of their own. To be human means to be self-moving. At the heart of this theory is the fool's propensity to rush about satisfying every desire, as Angiola did as a child. Because of their corrupted nature, the body-men live under the tyranny of self-love, "which compels them to make private utility their chief guide. Seeking everything useful for themselves and nothing for their companions, they cannot bring their passions under control to direct them toward justice."[158]

The impetus to the birth of *conatus*, and thus of justice, in the primitive giants' "wholly corporeal imagination" was "the frightful thought of some divinity." The precise circumstances that led to the recognition of *conatus*, and thus of human free will applied to the motion of the passions, involved copulation practices. Some two hundred years after the universal deluge, when the earth was finally reduced to a state of dryness, "it could send up dry exhalations or matter igniting in the air to produce lightning." The sky first "fearfully rolled with thunder and flashed with lightning" when primitive man was still sunk in his senses. The violent impression produced by the thunderbolts sank deeply into the folds of his brain, no doubt, and stunned his animal spirits into stupefaction, or wonder, characteristic of the vulgar ignorant "for whom everything is wonderful."[159]

A few of the most robust giants, scattered throughout "the forests on the mountain heights where the strongest beasts have their dens were frightened and astonished by the great effect whose cause they did not know, and raised their eyes and became aware of the sky." Because ignorance of cause leads humans to attribute their own nature to the effect, "and because in that state their nature was that of men all robust bodily strength, who expressed their very violent passions by shouting and grumbling, they pictured the sky to themselves as a great animated body"—Jove, who was trying to tell them something with the language of the "hiss of his bolts and the clap of his thunder."[160]

In one act of imagination the robust giants created both the divinities and themselves as human beings. Henceforth the language of the sky would be read through divination. At the same time, they conceived shame for their promiscuous ways: "So it came about that each of them would drag one woman into his cave and would keep her there in perpetual company for the duration of their lives. Thus the act of human love was performed under cover, in hiding, that is to say, in shame; and they began to feel that sense of shame which Socrates described as the color of virtue." This is what Vico means by saying that moral virtue began, as it must, from *conatus*. "By restraining their bestial lust from finding its satisfaction in the sight of heaven, of which they had a mortal terror," the giants began to exercise control over their passions: "they had to hold in *conatus* the impetus of the bodily motion of lust." "Thus they began to use human liberty, which consists in holding in check the motions of concupiscence and giving them another direction; for since this liberty does not come from the body, whence comes the concupiscence, it must come from the mind and is therefore properly human."[161]

Marriage, a union consecrated under the auspices of Jove, was the second human institution that arose out of the first institution, religious piety. Burial of the dead and, along with it, the notion of property followed. The rest of human history is a progressive education, in the sense of the Latin *educere* and *educare*: a process of drawing human form out of corporeal matter. "For heroic education began to bring forth in a certain way the form of the human soul, which had been completely submerged in the huge bodies of the giants, and began likewise to bring forth the form of the human body itself in its just dimensions from the disproportionate giant bodies."[162]

Vico stresses that humans made themselves out of themselves "The founders of gentile humanity in a certain sense generated and produced in themselves the proper human form." They did this "by means of frightful religions and terrible paternal powers and sacred ablutions." The process was a twofold education, of the spirit and of the body: "They brought forth from their giant bodies the form of our just corporature, and . . . by discipline of their household economy they brought forth from their bestial minds the form of our human mind."[163]

In the "Practic of the New Science," added as a conclusion in the third set of corrections and amendments in 1731, Vico ends his contemplative work with a practical addendum. All sciences that deal with matters that depend on human choice are "active." Therefore, the *New Science*, which concerns a matter that depends on human choice, should give indications for an applied practice. This "practic" is designed to delay, if not to prevent "the ruin of nations in decay," begins Vico. He ends it with these words:

> By these principles of metaphysics brought down into physics and then through morality carried into family government and thus into their own education, let the young be guided into good politics, and with their minds so disposed let them finally move on into jurisprudence. . . . And the youths will thus be brought to the true crossroads

of Hercules, who founded all the gentiles. Namely, whether they will take the road of pleasure, with baseness, scorn, and slavery for them and for their nations, or the road of virtue, with honor, glory, and happiness.[164]

Hercules is a poetic character that signifies the fathers of the first gentes, the first men in history who cultivated the land. The character of Hercules signifies that the first fathers were (1) just, (2) prudent, (3) temperate, (4) strong men. The tales of Hercules' labors are all stories about subduing the land, figured as a dragon: "They imagined the earth in the aspect of a great dragon, covered with scales and spines (the thorns and briers), bearing wings (for the lands belonged to the heroes), always awake and vigilant (thickly grown in every direction)." The earth was also imagined as a hydra, "which when any of its heads were cut off, always grew others in their place," and finally, as "a most powerful beast, the Nemean lion, which philologists hold to have been a monstrous serpent. All these beasts vomit forth fire, which is the fire set to the forest by Hercules."[165] Hercules was the slayer of monsters in two senses: he cultivated the land, and he cultivated his humanity. Vico is telling the youth of his day to cultivate their own heroic virtue, as Angiola Cimini did, through piety and the study of human letters. Theirs is the choice to become fools or gods: to slay the monster within or to become one.

Conclusion: Hercules and Mediocrity

Hercules is a rich, ambivalent figure to propose as a role model for young adults. In Baroque iconography he is often depicted, club in hand, at the crossroads between virtue and vice, clothed in the skin of the Nemean lion. This terrifying monster was no ordinary lion. It was said to be either one of the monstrous offspring of Orthus and Echidna, or of Typhon, or else it had been suckled by the moon goddess. Since it was invulnerable to weapons, Hercules had to strangle it with his bare hands. The returning hero skinned it and dressed himself in its pelt, with the scalp serving as a sort of hood, which is why in many classical depictions Hercules' head emerges from the beast's mouth. In Porta's physiognomy Hercules is used to illustrate heroic virtue. Figure 16 shows the well-proportioned, naked body of the hero, who rests on his club; poking out from under his arm is the lion's head, whose profile is drawn to emphasize similarities with Hercules' own profile.

The tales of Hercules' labors all involve battles with various monsters, animals, and gods: his opponents include the Hydra, a hundred-headed snake; Geryon, the three-headed giant; the man-eating Stymphalion

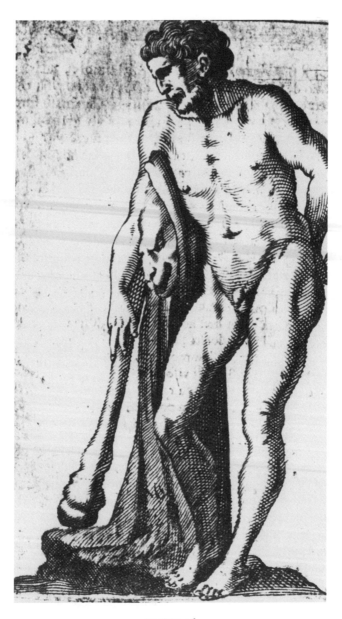

FIGURE 16
Hercules resting on his club, with the
head of the Nemean lion under his arm.
From Porta's *Della Fisonomia dell'Huomo,* 1652 ed., p. 512.
(Courtesy of the Bancroft Library)

birds; Cerberus, the triple-headed hound at the gates of Hades; and count-less others.[166] In some versions, in order to retrieve the flesh-eating mares of Diomedes, he was forced to sleep with the queen of Forest, whose body was that of a snake from the waist down. He gave the she-monster three children before being permitted to leave. The rest of the tales narrate a spirited campaign of vengeance systematically conducted against his en-emies. Mired in dung, bathed in the blood of monsters, gods, and animals, the hero is flagrantly forced to mix, merge, beget with monstrous figures, some created by his own mother, Hera. He is continually compromised in his humanity. The series of battles with monsters, the progeny of promis-cuous couplings between humans, gods, and animals, and his encounters with man-eating animals trace a cycle of birthings, killings, ingestions, and defecations that persistently invoke the precarious bodily confines that divide man from monster.

In many ways, Hercules was only half man. His other half was beastly, not godly, and thus he typified the heroic spirit for Vico. While Hercules was indeed "almost the ideal embodiment of the Greek settler, who de-stroyed aboriginal monsters and gave peace to the regions which he tra-versed," he also had affinities with the monsters he encountered. "He was the cleanser, the hero who could not really be tolerated within the city be-cause of his bestiality, his crudity, his violence."[167] His ambivalent nature, characteristic of a *pharmakon,* a cure that kills, is what makes him such an apt role model for Vico. Hercules' series of encounters with monstrous gods and animals was what forced him to summon up heroic exertions in order to combat the figures of human alterity that doggedly besieged him. Hercules is a good role model for those who have yet to conquer their own humanity by dominating the savage natures that dwell within.

This interpretation is confirmed by an emblem in Cesare Ripa's *Ico-nologia* under the rubric of "Heroic Virtue." In Rome, writes Ripa, there is a statue in the Campidoglio that figures Hercules clothed in the lion's hide with a key and holding three golden apples in his left hand. The apples signify the three heroic virtues attributed to Hercules: the first is modera-tion of rage; the second, temperance of greed; and the third is disdain for delights and pleasures. Heroic virtue, Ripa explains, is when reason has subdued the affects of the senses.

In another emblem entitled "Dominion over Oneself" a man is "seated on a Lion, that has a bridle in mouth, & he controls this bridle with one hand, & with the other he pricks this Lion with a spur . . . in order to show that reason must keep the spirit [*animo*] in check when it is too impudent, and prick it when it proves tardy and sleepy." The bridled lion also ap-pears, this time controlled by a woman, under the rubric of ethics: "Ethics signifies the doctrine of behaviors, including the concupiscent and irascible appetites. She holds close to her the Lion, a noble & ferocious animal, bridled, to signify that she restrains the animal part of man." In her left hand the dominatrix holds up a plumb rule or a mason's level: "This doctrine thus teaches man that the sensual appetite conforms to the rectitude & evenness of reason when it leans not toward extremes, but maintains itself in the middle."[168]

Mediocritas, maintaining the middle ground between reason and desire, tempering the bestial and the divine in order to produce a harmonious human blend, is the ideal of early modern ethics, the science of behaviors and passions. This ideal is expressed in the notion of "justice," at once a juridical, physiological, and ethical concept. The key to "humanness" lies in the ability to combine into a just whole the opposing natures of the divine and the bestial. It is a doing, a making, a manufacture.

According to the authors we have read, we humans make our selves in the same way we make our pet monsters: in a laboratory according to a recipe of natural magic; in a grotto according to the laws of hydraulics; on the dissecting table according to the latest physiological theory. With our minds, our passions, and our bodies, say the early modern natural philosophers, we manufacture ourselves and our creations in our own image. The descriptive, manipulative, coercive, and transformational modes of monster-making that I have discussed in previous chapters are all employed when it comes to philosophies on the making of man. Vico's blending of physiognomy with Borelli's application to the human body of Galilean physics provided an anachronistic detour around Cartesian dualism that was destined to be rejected by Enlightenment philosophers.

In general, we may observe that monstrosity tends to be generated from overly strident definitions of matter and form. The source for these monster-making definitions is the oft-repeated Aristotelian dictum that natural bodies have an innate motive force, a formative virtue, while artificial, and hence, unnatural bodies are passive matter, moved by an external force. Whenever humans define their essence as a hybrid of passive, material body and active, conscious mind, *res extensa* and *res cogitans,* monstrous images of self are bound to ensue. As long as human beings erect their identity on the foundation of an uneasy dichotomy between mind and body, pitting one imagined aspect of self against another, the ensuing hybrid creature is bound to be monstrous.

As we shall see in the next chapter, the struggle of the mind to overcome the body's unruly appetite for pleasure can result in a sort of paralysis of the machine. Ulysses bound to the mast, unable to respond in action to the allurements of his senses, becomes the emblematic figure for industrial, bourgeois man, chained to his machines in exchange for freedom from labor.

SIX

Monstrous Metaphor

Our search for monsters has led us from the garden into the natural philosopher's laboratory, through the museum hall and past its moving statues, into our own backyards and up onto the anatomist's dissection table. We have discovered the monster in the deep recesses of our hearts, catching glimpses of the animal other in distorted mirror images as we pass through the physiognomist's study. We have discovered the monster in our own bodies, transmogrified by vice and rehumanized by virtue, then cast in the gigantic lineaments of the body politic. The monster is our primitive past, it is our barbaric future. Neither real nor imaginary, neither a symbol nor a thing, the monster continuously asserts its presence by its powerful force on the imagination.

What we will discover in this chapter is that stupor and marvel—the same emotions that emerge when monsters are put on display and when automatons execute their mechanical performances—were in themselves monstrous affects. The art of inducing wonder or admiration was considered the basis of a pedagogically sound approach to learning, but if taken too far, or if sought after purely for its own sake, it was thought to bring with it dire consequences. As we shall see, there is more than one way to generate monsters and to lose one's "human" faculties. Tinkering alone in one's laboratory or gazing at one's image in the undulating surface of a mirror was a risky pastime, just as allowing one's imagination to wander could bring unwelcome results for an expectant mother.

But as Gustave Le Bon noted in 1895 regarding the psychology of crowds, the social context of a group could be just as dangerous as the solitude of a museum. Under the spell of a charismatic leader, a hapless member of a crowd can lose his "conscious personality," becoming prey to the suggestions of the leader, and allowing his "unconscious personality" to come to the fore. Bereft of the resistance afforded by the conscious mind, the leader's suggestions immediately transform into acts, and "these, we see, are the principal characteristics of the individual forming part of a group." The figure Le Bon uses to describe this feckless and diminished individual who has fallen under the monstrous spell of a leader is apt: "He is no longer himself, but has become an automaton who has ceased to be guided by his will."[1]

Emblematic Wonder

The emblem was a popular artifact combining a title, an epigram, and a visual representation, designed for didactic as well as for entertainment and decorative purposes. Now, unlike the more "heroic" *imprese*, which, because of their function to nobilitate family coats of arms and so forth, were not supposed to make use of fantastic, monstrous, or artificial bodies, the emblem "allows a plurality of historical, or fabulous, or artificial, or natural, or chimerical Figures"; in fact, the true emblem "rejects human Bodies in its figures."[2] The most famous emblem book was Andrea Alciati's, with over 175 editions in every European language along with numerous commentaries and imitators.[3] One of his emblems makes use of a popular monster we have already come across in other contexts: the siren.

> Birds without wings, and girls without a mouth,
> and fish without a mouth they nevertheless sing with their lips.
> Who would think that any such creatures exist? Nature denied
> that these things can be combined;
> but the Sirens show that it is possible.
> She whose form ends in a black fish-tail is a woman of seductions,
> because lust carries with itself many monsters.
> Men are attracted by appearance, by words, and by brilliance of
> spirit,
> that is, by Parthenope, by Ligia, and by Leucosia.
> The Muses tear off their feathers, and Ulysses mocks them.
> That is to say, scholars have nothing to do with a harlot.[4]

Emblems were distinguished from their more erudite counterpart—the impresa—precisely for their popular content: the message was intended not so much to provide a stimulating brainteaser (as did another more ludic form based on the late-Renaissance vogue for hieroglyphs, the rebus sonnet),[5] but rather to illustrate an imperative directed to the moderately educated class of people, those who "delle lettere humane sono mediocremente infarinati." Emanuele Tesauro gives the clearest and most succinct definition: "A Popular Symbol; composed of Figure and Words, signifying by way of an Argument, some Document belonging to human life: & therefore, displayed for decoration & ornament in Paintings, in Halls, in Theatrical Sets, in Academies; or printed in books with Images and explanations for public teaching of the People."[6] Monsters were at home in these moralizing picture books because they served to provoke not horror, as they would in the nineteenth century, but quite another sort of affect, that is, the didactic passion par excellence: *meraviglia* or wonder.[7]

All Aristotelian poetics (and *seicento* theories of wit are almost exclusively derived from Aristotle's *Rhetoric* and *Poetics*, heavily mediated by Horace) are based on the fundamental assumption that "it is owing to their wonder that men both now begin and at first began to philosophize."[8] The notion of *thauma*, or wonder, as a nursemaid to learning was a Platonic notion as well. As Socrates explains to Theaetetus: "This sense of wonder is the mark of the philosopher. Philosophy indeed has no other origin, and he was a good genealogist who made Iris the daughter of Thaumas."[9] Iris, the personification of the rainbow, was also the messenger of the gods; hence, through their sense of wonder, humans have access to the other world of ideal forms. Our curiosity is a divine attribute.

Aristotle not only sees wonder as a stimulus to metaphysical endeavors, he also paradoxically describes our capacity for wonder, and the desire for learning that it entails, as the characteristic that defines us as rational animals. Learning grants humans access to their most "natural" state: "Wondering implies the desire of learning, so that the object of wonder is an object of desire; while in learning one is brought into one's natural condition."[10] The link between desire, learning, and wonder remained a crucial one. Sforza Pallavicino's formulation can serve as a seventeenth-century exemplar, in which delight (*diletto*) has broadened into a broader notion of pleasure (*piacere, contezza*): "Wonder is the fountainhead of an ultimate intellectual pleasure, inasmuch as it is always linked with knowing that which was previously unknown. And the more the thing was unknown or even contrary to our beliefs, so much the more is the wonder and, along with it, the pleasure of having acquired a familiarity that up until that moment had seemed remote and hopeless."[11]

Wonder, *admiratio, meraviglia*, or in French, *l'admiration*, was primarily visually stimulated.[12] Aristotle teaches that antithesis, or artful paradox, is what "impresses the new idea more firmly"; and "liveliness," the most important quality to effective speech "is got by being graphic (i.e., making your hearers *see* things)."[13] It is not surprising, then, in a time "in which the tendency to images reached its climax," as Mario Praz puts it, that the method of teaching through images was also launched, by

J. A. Comenius in his *Orbis Sensualium Pictus* of 1658.[14] Campanella's visual encyclopedia, built into the seven circular walls of the City of the Sun, was an ingenious elaboration of this antibook, antischolastic pedagogy based on the instant visual grasp of material.[15]

Emblems, as a hybridized genre combining a motto, a picture, and an epigram—which the delighted mind grasped as one, divining connections and correspondences between the parts—were a new and wondrous sort of sophisticated pedagogical tool. And the emblem form itself, apart from its fantastical content, had an analogous structure to a monstrous hybrid, instilling wonder and thus predisposing spectators from the outset to learn and delight in their learning. Just as in imagined theaters of memory, in which unusual figures were used to better "fix" the sequence and contents of thoughts, the fantastical, chimerical bodies represented in emblems aided in an effort to induce in the emblem reader an altered state of cognition, one provoked by apprehension of the rare and extraordinary, precisely the affect provoked by monsters.[16]

Excessive Wonder
"Ne Peut Jamais Être Que Mauvais"

Wonder was codified as a good affect that could aid cognition by leading the subject into a perceptually altered state of consciousness, one which could greatly facilitate the learning process. The implications of this pedagogical approach were, of course, hardly lost on the church in its efforts to persuade erring pilgrims back into its cathedrals. Visually, the effort to artificially induce *meraviglia* led to the stupendous church interiors that typify Baroque Rome: all the clichés that we identify with the Baroque aesthetic, trompe l'oeil techniques, the dome opening up into the heavens, painted figures merging into stucco, and stucco into columns, theatrical stagings, the *quarantore* Lent extravaganzas engineered by the best artists of the day. Art history has explored this connection very thoroughly, along with the concomitant theatricalization of the public space.[17] The point to keep in mind is that, as we shall see, it was primarily Jesuit orators who were responsible for transposing this notion of instant visual apprehension into literary usage as instant apprehension of concepts through the wondrous medium of metaphor; for example, Matteo Peregrini suggested that *acutezze* "be used to make many things, or important ones, appear to the sight of the listener in a few words."[18]

But this positive codification undergoes at first a subtle and later not so subtle change by the last quarter of the century. The general swing toward classicism that came in part as a reaction to rococo tastes already finds its theoretical formulation in Descartes's 1649 treatment of wonder in *Les Passions de l'âme*. His overall project is one of reducing "the machine of our body" into its constituent parts, so that division of labor is clearly demarcated between soul and body, the soul being left with nothing other than thoughts, those that derive from outside stimulation and those that well up solipsistically from inside.

In spite of Descartes's claim to utter originality in setting up his taxonomy of the movements of the soul, *l'admiration* is designated as "the first of all the passions" serving man to learn in both the Aristotelian and Cartesian schemes: "And one may say in particular about admiration, that it is useful because it makes us learn and retain in our memory things of which we were previously ignorant. For we only admire what appears to us as rare and extraordinary: and nothing can appear in this light unless we were ignorant of it, or even because it is different from the things we already knew: for it is this difference that causes us to call it extraordinary."

The province of admiration is the "rare and extraordinary." It follows that a capacity for wonder is a boon for acquiring knowledge, since it makes us apprehend and retain new things, "and so it is that those who have no natural inclination for this passion are ordinarily quite ignorant." So it is good to be born with an inclination to wonder, "because this disposes us to acquiring the sciences."

This positive valorization, however, has its limits. The moment one goes too far, a serious threat to reason looms on the horizon. One has strayed into the regions of *l'étonnement*. Astonishment is "an excess of admiration that can never be anything but bad [un excès d'admiration, qui ne peut jamais être que mauvais]." The surprise provoked by astonishment is so violent that the animal spirits in the brain are so occupied in maintaining the impression of the new object being admired that none flows down into the muscles, "which makes the whole body remain immobile like a statue, and one can only perceive the first side of an object that is presented, nor is one able to subsequently acquire a more detailed knowledge of it." Not only does excess wonder result in total seizure of the machinery of the body, blocking the possibility for further perception, "this could entirely remove or pervert the use of reason."[19] Reason perverted, reason removed, a man without his reason is nothing more than a beast (Orlando "furioso"), or at best, according to the mechanistic philosophy, a machine.

Tesauro, who is much more at ease with these immoderate states of being, writes of just such a paralyzing effect due to excessive wonder. He is discussing automatons, in fact, described by Cardano: "those ivory Dolls, who dance about in turns on their own, and move their feet, arms and eyes according to the rhythm of the beat." The effect of astonishment on the spectator-subjects was such as to turn them into doubles of the objects they admired, and conversely, the objects into spectators: "so much stupor on the part of the onlookers that they appear to be statues; & the statues appear to be onlookers."[20] The danger hinted at by Descartes and described as a physiological immobility that leaves the body "like a statue [comme une statue]" is here more explicitly revealed as a collapse of distinction between subject of consciousness and object of apprehension.

Wonder acts as a sort of trance state, an intermediate state of being in which the ego boundary is more permeable; consciousness of self fades to allow more intense and memorable imprinting of the object onto one's cognitive and memory organs. This is the positively connoted aspect of wonder. But, as in all altered states, there is a danger of remaining captivated by and in the "other," so that self runs the risk of being annihilated.

In the terms of early modern thinking, as we have seen, human authenticity is based on the proper functioning of reason; if that capacity becomes impaired, the person deprived of reason, *il forsennato*, is transformed either into a mad beast or an automaton. The sole difference between a live human being and a dead one, explains Descartes, is the same as that between a wound-up watch and one that has run down: "The body of a living man differs as much from that of a dead man as a watch or another automaton (that is to say, any machine that moves of its own accord), when it is wound up and it has the corporeal principle of movement within itself for which it was instituted, with all that is required for its action, and the same watch or other machine, when it is broken and its principle of movement ceases to act."[21]

Ultimately, then, wonder is a double-edged sword, and "we must . . . strive to deliver ourselves of it as much as is possible."[22] The remedy is clear: it consists in acquiring knowledge of many things, "and in exercising oneself in considering all things that may seem most rare and most strange." One would think that habit would dull the senses and protect those vulnerable to astonishment by making them blasé. It is true that "this passion seems to diminish through usage, the cause being that the more one encounters rare things that one admires, the more one becomes accustomed to not admiring them," but even the most conscientious seeker of rarities can fall prey to a kind of sickness of curiosity.

The side effect of this "maladie" is a sort of blindness that makes every object equally fascinating. One's cognitive judgment is impaired, so that one falls into a pernicious habit that "disposes the soul to stopping in the same fashion on all the other objects that are presented to it, just so long as they seem even slightly novel." Those who are afflicted with this illness, that is, "those who are blindly curious," find it difficult to recover, precisely because they seek out "rarities only to admire them, and hardly in order to understand them." Gradually the immobility that affects their cognitive capacity becomes a fixation, they are unable to leave their state of astonishment, and all objects, regardless of whether they are truly worthy of wonder or not, have the power to captivate them "because they become gradually so full of wonder that even things of no importance are just as able to stop them as are things that are more useful to research."[23]

The key to this ethics of curiosity clearly lies in the distinction between seeking rarities in order to know them better (the Baconian project for the new experimental science),[24] and searching out oddities purely to admire them (the cheap thrill behind curiosity collections). In other words, the technological, use-oriented motivation is perfectly acceptable in the new scientific ethos; but pleasure seeking, using the rare and extraordinary solely as a kind of drug that induces an egoless state of wonder is a highly dangerous and even immoral pastime. Or as Francis Bacon (1561–1626) expressed it: "Knowledge that tendeth but to satisfaction is but as a courtesan, which is for pleasure and not for fruit or generation."[25]

The Debate over the Use of Witticisms

Descartes was not alone in decrying the inordinate appetite of many moderns for astonishment and wonder. Tommaso Stigliani complains that "at one time readers were satisfied with a not-so-bad reading, then they wanted excellence, next they desired wonders, and today they seek stupefactions, but after finding them, even these are tiresome and they aspire to be flabbergasted and stunned."[26] Stigliani characterizes the contemporary literary sensibility as "overfed, so very satiated and so very jaded" that the reading public is like a pregnant woman, prey to absurd cravings, whose desire degenerates into a faulty wish for filth.[27] And while Sforza Pallavicino reiterated in 1662 that "the principle delight of the intellect consists in wonderment," Matteo Peregrini had already complained in 1639 that "instead of delight, Witticisms, because of their overly elaborate conditions, can give birth to nausea and indignation."[28]

The terms of the debate over poetics are clear: on the one hand, those like Peregrini, Sforza Pallavicino, and Daniello Bartoli who believe in a regulated use of the "pellegrino"[29] or extraordinary (the so-called moderate Baroque critics); and on the other hand, radical supporters of the Marino school, like Tesauro, who believe that "the nausea of everyday things" must be relieved no matter how absurd the lengths to induce a state of "meraviglia."[30] In both cases, however, illness, nausea, and satiety supply the terms for the argument. Bartoli describes the modern parlance "il dir moderno" as being so full of variety, putting "such a copious amount of delicious foods in front of you that, snatching them away the instant you taste them and putting other new ones in their place, you are kept constantly satiated and always hungry, in line with the ancient law regarding the most noble banquets." But when the author is guilty of poor judgment, the banquet of conceits turns into a sickening display of empty affects. Confesses Bartoli, "For myself, when I happen to hear affects managed in similarly inappropriate fashions, I feel more nausea than someone suffering from seasickness."[31]

The issue at stake in what at times turns into acrimonious, bloody conflicts is more than a matter of theoretical speculation between pedants. If it was becoming more and more difficult to satisfy the heightened taste for the new, rare, and unusual—what art historians refer to as the Mannerist penchant—perhaps this hunger for the grotesque, the deformed, and the "smisurato" (or "out of proportion") had run its course. Traditional Aristotelian schemas were clearly proving insufficient and even outmoded. *Seicento* theories of poetics elaborate one of the last great attempts to construct a normative literary etiquette based on taxonomy: genus, species, essence, attribute, and differentia, shuffled, stretched, redefined, until the rubrics ceased to maintain their distinctiveness, and thus their power to regulate. In the eighteenth century, poetics as a prescriptive discourse, as a rhetoric of literary manners, transformed for the first time into aesthetics; art was relegated to the science of the beautiful. Its pragmatic function withered away; art became mere contemplation.

But at this point Stigliani senses that precisely what is endangered by this exasperated search for the monstrous is the link with the didactic purpose that originally justified such a search. The monstrous was formerly admitted as a means; it had now become an end. The need to regulate taste concealed a more urgent need to safeguard the ethical basis of allegories, fables, and fiction in general: the ancients' and Christian doctrine of veiling the truth either to make it more palatable to the ignorant or to hide its mystery from those who were unworthy to receive it.

In short, a hierarchy of taste implied a hierarchy in the social sphere. If there was no way of distinguishing between good conceits and inane or ridiculous ones, then absolutely anyone could try her or his hand at composing them and there was no end to what might follow. What other distinctions could begin to crumble? The use of *acutezze*, Peregrini repeats insistently, "should be made not in a poor manner with absolutely no norms, as many are wont to do; but with moderation, norms, and reason." It is a question of a general attitude, that of moderation, and it consists in knowing how to take the middle path, which "in moral thought and in that of eloquence, holds the place of virtue." Virtue, in ethics as in rhetorics, is a matter of appropriateness: "The norm for such a practice would be first to choose the suitable materials . . . then temper the quality and quantity . . . according to the general rules of the art, principally according to those of the Appropriate."[32]

"Among the traits that distinguish an Orator from a Fool," observes Peregrini, "one is the observance of Decorum." This reasoned plea for a regulated usage of the "mirabile" only barely disguises the Jesuit's real panic: that any kitchen drudge or dogcatcher will try his hand at literary compositions and get it into his head that he, too, deserves a good position in the academy: "So much the more because if these kinds of compositions do not strictly require excellence of judgment to block them out or build them, since they do not bear within them the need for any other particular branch of learning, I don't know why a Cook or a Dog-catcher couldn't pick up a pen, too, and demand some nice position in the Academy."

But Peregrini's snobbism is a complex matter. To begin with, in his own estimation, anyone who would dedicate his time to studying "le Acutezze mirabili"—the subject of his treatise *Delle Acutezze*—is highly suspect. What do they expect to gain from their endeavor, anyway? "Maybe to grandly entertain others? This is the work of Actors, Conjurers, Parasites, Flatterers, & other vile and servile people like them," he spits out. Those engaged in making their living through entertaining others are the lowest of the low, along with "Prostitutes and their ilk, competing in Witticisms." Perhaps that is why his publisher felt the need to assure readers who made it through the 256 pages, ten classes of "Acutezze Vitiose [Defective Witticisms]," and twenty-five "Cautele per l'uso delle Acutezze [Precautions in the use of Witticisms]" that the work was written by Peregrini "principally for his own recreation, while he was composing the most weighty Little Work of *Political maxims*, in Genova at the Palace of Fassuolo belonging to the Most Excellent Prince Glo. Andrea Doria, his Lord."

Politics is the stuff of weighty intellects; literary conceits are recreational trifles. If the study of *acutezze* "are to be esteemed as belonging to empty, feather-weight intellects," those "lacking in judgment" and "swindlers," then the very presence of the treatise is somewhat of a puzzling embarrassment to the author, to be sure.[33] But there must be something more important involved to have merited such a sustained effort. Clearly, even the author is ambivalent about his participation in the activity.

Peregrini goes so far as to discourage any would-be writers from burdening the world with their mediocre compositions. In fact, those who are insufficiently schooled in acute wit are advised that "the world will be able to go on without your Epigrams, Madrigals and Sonnets." If the only reason they attempt to publish is for enjoyment or glory, then his contempt knows no limits: "Abortions, monsters, insipid Things, or otherwise useless, and mere pieces of junk; the World, in my opinion, has so much of these things, that anyone who taxes himself by adding to them can only be very importune and greatly indiscreet." One cannot help feeling that the author "doth protest too much."[34]

But we mustn't be too hard on Peregrini. He, after all, is only too aware of the beam in his own eye. As a preacher, he knows well the temptation to succumb to affectation. Because of the nature of witticisms, "a far too alluring matter, and therefore most powerful in confusedly dragging one's wits behind it," even those who deal with "holy matters" end up being contaminated with the same vile vice. Alas, as Peregrini admits in relation to humorous witticisms (but I think his attitude can apply generally to the use of all *acutezze*): "We want to please, we want to delight, we want to make ourselves admired; O fools that we are."[35] Like Tesauro's spectators who turn into the wondrous statues they admire, like the literary vulture who stole the sayings of the small-headed siren only to become entrapped in her net of words, Peregrini not only wants to provoke wonder in his auditors, he wants to *be* admired, he wants to usurp the place of the object of admiration. The rare and unusual, like the monstrous in general, has a propensity to turn back onto its user, infecting, contaminating, breaking down hierarchies, disintegrating ethical and moral constraints, blinding and nauseating those who are temerarious enough to use it.

Ten years before Descartes's formulation, Peregrini described the physiological effect of *admiratio* as a blinding force that is so powerful it disables the better judgment of those who fall under its spell: "The ray of novelty is at first encounter a blinding ray, even to the most acute visions: and its allurement is so powerful that while others fail to heed the warnings, it easily puts to sleep the use of good judgment. Because rarity has such a great part in Witticisms, novelty necessarily also has a large place in them; and even more in defective ones."[36] That is why the whole project of writing about the new, rare, and unusual is so fraught with ambivalence. "The more wondrous is the wondrous, the more it has of appearance and less of substance."[37]

The publisher Giovanni Battista Ferroni succinctly expresses the risk (and Peregrini's strategy for avoiding it) in the misogynist terms that were fashionable in the *seicento:* "Eloquence . . . also has her Lusts; but they are Lusts without lasciviousness. She is hardly transported into the incontinence of talking about it, which is why it is no wonder if she is not attractive to libidinous intellects; because the elegance of this Author has beauties up for marriage, not for prostitution."[38] Appearance over substance, prostitution versus marriage, wanting to please and be admired, all signs point to the alluring nature of this "overly enticing matter": it tends to corrupt the moral standards of its listeners, it draws them into the desire to let go of all inhibitions, to lapse into the purely enjoyable, into that state of wonder in which reason becomes impaired, in which the purely corporeal, whether that be figured as bestial, mad, or machine-like, takes over.

"Lovers," avows Peregrini, "are, in a certain sense, out of their minds; that's why baby talk is not at all inappropriate for them." Lovers, children, "the popular mob," and fools are all more susceptible to the allurement of the *mirabile*. Naturally, so are the insane. And anyone who persists in using inappropriately ridiculous metaphors runs the risk "not only to have himself ridiculed, but if he goes one much more, to have himself considered absolutely insane." How more explicit can Peregrini be? As he warned earlier, this is "an overly enticing matter, and therefore it is very powerful in confusedly dragging one's wits behind it."[39]

In some ways, in spite of his greater fear, or perhaps because of it, Peregrini has a more sophisticated understanding of the ambivalent working of wonder than both his conservative and radical counterparts. The most striking aspect of Peregrini's treatise (on what are called variously "acutezze," "argutezze," "spiriti," "vivezze," "concetti," "acumi," "conceits," "witticisms," "agudeza")[40]—unquestionably the first of its kind, predating Gracián's *Tratado de Agudeza y Ingenio* and Tesauro's *Cannocchiale Aristotelico*—is its understanding of the mediated, social nature of wonder.[41]

Peregrini, consonant with the rhetorical tradition, repeatedly draws attention to the fact that the *use* of a particular conceit and not its *content* is the best criterion for distinguishing between a tasteful and ridiculous witticism. In discussing defects that can mar witticisms, he remarks that in general "Vice touches rather the use than the substance." Conversely, "it is not possible to demonstrate precisely . . . with a determinate line" between good conceit and a bad one through purely formal means. "Once again," he must "remind us of the difficulty of the subject we have undertaken." The brilliance of particular witticisms are clear, but "how this comes about is very difficult to explain," he admits repeatedly. Even their classification is beyond his more than adequate intellectual capacities: "It is so difficult to distinguish the genera and species of Witticisms," he laments, thus rejecting the possibility for an objective science of wit. *Delle Acutezze* is in fact a kind of speech-act theory of rhetorical figures in which the speaker's intention and the effect of the witticism on the hearer both constitute the ends of the act and provide a means of classification:

"Certainly," he reasons, "the intention of the person who speaks has a great part in all Witticisms"; "in fact, according to . . . Cicero the intention of the Speaker alone can make the Saying playful or not."[42]

By privileging the final cause of witticisms ("the end . . . to make oneself admired")[43] instead of their formal cause, taxonomy yields to psychology. At all times, wonder functions not as an affect inspired by the particular witticism in the hearer's brain, but as a tripartite act involving "il Dicitore," the speaker, "l'Ascoltante," the listener, and the "ingegno" or ingeniousness[44] of the speaker. The listener never admires the witticism itself. He or she admires the *ingegno* that produced the witticism. Delight and marvel ensue not from the object of wit but precisely from the speaker's wondrous talent to construct the witticism: in short, the speaker becomes the object of wonder.

This essentially mimetic, mediated nature of wonder is explained in chapter 11 of his treatise, in which the *Acutezza mirabile* is given its definitive formulation: "An artificial enthymematic linking of several things in one Saying in such a way that one falls so extraordinarily fittingly beside another, that the ingeniousness of the Speaker becomes the object of admiration, by which the Listener is most forcefully amused." To put it another way, the art of *acutezza* lies in knowing how to figuratively link together parts with "such a rare fitting," but what provokes wonder is not the result of this extraordinary arrangement; rather, the rarity must be such that "the power of the ingeniousness makes itself the principal object of admiration." "To delight, just as much to show off one's wit, are effects to make the Speaker amiable, which in similar occasions touches on his supreme goal."[45] The new, rare, and unusual once again have power to turn their user into the "principal object of admiration": whence lie their power, and their danger.

All varieties of defective witticisms, whether they are "Fredde, Stiracchiate, Fanciullesche, Vuote, Insipide, Inette, Stolte, Niquitose, Sfacciate [cold, drawn out, childish, empty, insipid, inept, foolish, harmful, impudent]," or "Buffonesche [doltish]" partake of a common characteristic: they go too far beyond the limits of social decorum. They are either "immoderate" or too patently a lie; they "lapse into the Improper" by straying too far from the common in search of greater rarity. Others simply demonstrate "a manifestly great lack of judgment," or "they convey a sentiment contrary to the rules of moral virtue," or they are clearly "dishonest."[46]

In every case, in all cases, whatever their particular sin, Peregrini characterizes this immoral or uncivil use of figures of speech as "il *transandamento*," literally, a going beyond. The alluring or enticing power of tropes is precisely a call away from the virtues of social moderation and civil behavior, into a "beyond" that is akin to the state of mind that lovers, children, madmen, and beasts inhabit. They induce a state of powerless pleasure in their listeners, and even more in their users. They beckon the courtier away from the highly sophisticated but highly regulated practice of "la civil conversatione" into antisocial behaviors displayed "in reading those modern books of ours that on every page have continual skirmishes between these fanciful locusts."[47] Here, the violent, rivalrous nature of trafficking in witticisms is made explicit: the exchange tends to mutate into a rivalrous display of wit between practitioners that degenerates easily into "skirmishes."

Matteo Peregrini was a founder of the Accademia della Notte, a promoter of many others like the Accademia de' Gelati, a doctor of theology, philosophy, and law, and a "Learned Archivist, Writer of many works," such as *Il Savio in Corte* [The Wise Man at Court] (1624) and *Della pratica comune a' prencipi e servidori loro* [On the practices common to princes and their servants] (1634).[48] The reason why he is so concerned about the improper usage of metaphor is because in his experience, and according to the rhetorical tradition that he subscribes to, the game of wonder always takes place in a social context; it is always directed by aims of persuasion. What is at stake is not just ways of writing poetry or sermons but rather ways of affecting people in the social realm, either from the pulpit or in courts, where brilliance of wit had become cause of unrelenting rivalry, resulting in the making and breaking of the prince's favorites and their accompanying stipends.

Wonder, used as a game between rivals, as a weapon to stupefy and manipulate hearers, was a dangerous instrument indeed. Even Lodovico Tesauro, a defender of Marino and a supporter of radical experimentation, perceived the dangers of exceeding "poetic measures," a tendency characteristic of ambitious emulators and would-be rivals of Marino: "Some writers, ambitious to have themselves honored by means of their arrogant caprices and strange conceits, have overpassed those poetic measures, which are usually obtained first by natural talent and then by practice through long and continuous study and which everyone knows Cav. Marino tempers very well within the terms of moderation; it is no wonder if they fall flat on their faces over the precipices, overflowing into excesses of immoderate, indiscreet, insolent hyperboles, and thereby forming such monstrous chimeras and portents."[49]

Previously we saw that monstrosity was as likely to be found in museum collections by looking in distorting mirrors as it was in preserved fetuses floating in glass jars. Baroque theories of wit reveal that the rare and extraordinary could be discovered not only in "found objects," New World pygmies, marine monsters, and basilisks, but just as readily in the everyday speech of men of letters.

Tesauro's Aristotelian Telescope *Sees Further*

Peregrini denounces the theoretical study of *acutezze mirabili* (wondrous witticisms), yet he himself participates in it; his ambivalence is manifest in his actions as well as at the thematic level of his text. But the ambivalence of wonder was encoded into the notion of *thauma* from its very beginnings. By silencing reason, thus suspending judgment, the true philosopher enters into the mysteries of the divine. Iris, the shimmering rainbow, carries messages between humans and gods. Wonder is very definitely a kind of *transandamento* (Plato). The object of wonder is an object of desire, the desire to learn, and in learning one is brought into one's natural condition as a reasoning human being. One silences reason in order to improve reason. Wonder is a "coming back" into the natural (Aristotle).

In the first understanding, wonder is an intermediate state that allows communion with the divine nature of the universe. In the second, wonder defines what makes us most human: our reasoning powers. As we have seen, the key to understanding the ethics of curiosity lies not in a phenomenological description of *thauma* but in its use. As a means to improved cognition, it enhances the human condition by provoking a state of mind that is conducive to learning and remembering. As an end in itself, it leads to antisocial, pleasure-seeking behavior.

When applied to the *seicento* passion for exasperated witticisms, the properly regulated usage of *il mirabile* became the focus of acrimonious debate. For Peregrini, the search for the rare and extraordinary aids us in revealing the divine harmony of things; a successful conceit is a well-proportioned, pleasing combination or arrangement of terms into a wondrous whole that shows off the extraordinary ingeniousness of the speaker. An immoderate use leads to rivalry, and reveals nothing but the bad side of stupor: the bestial, the infantile, the crazy, the antisocial. When the rare and extraordinary are sought out as an end in themselves, the search itself degenerates into a monstrous pastime.

Emanuele Tesauro's predicament is far less contradictory than Peregrini's dilemma. If we can characterize the conservative and moderate Baroque critics (Peregrini, Pallavicino, and, in most respects, Baltasar Gracián) as viewing the rare and extraordinary as a means to revealing divine harmony, the radical critic tended to see the rare and extraordinary rather as monstrous in themselves, and conversely, monstrosity as the very essential nature of the universe. Tesauro unambiguously embraces *meraviglia* as both a means and an end. In embracing the passion of wonder, side effect of the new, rare, and unusual, he also opens the gate to a proliferation of monstrosity. He puts both wonder and monstrosity in a central position.

For Peregrini the state of stupor is essentially amoral and asocial because (1) it impairs good judgment and reasoning; (2) it erodes the social hierarchy. For Tesauro the faculty of *arguzia* is the most properly human attribute; it is an aid to the civilizing process, what rescues us from lapsing into the bestial, as well as a vestige of the divine in our human souls. It is a "divine Offspring of the Intellect"; and the *argutezza* itself is a "most pleasing adornment to Civil conversation . . . vestige of the Divinity in the Human Soul." Not only does the art of composing *argutezze* cause the divine part of our human nature to resonate with God—the most witty and charming being of all—we should never be afraid of contamination from the ridiculous or sordid materials that Peregrini so despised, because "the ray of the human Intellect" is similar to that of the sun, which has the privilege of ranging over the impure things of this world while maintaining its own purity.[50]

Like Gracián, who believes that *agudeza* is our most sublime faculty, promoting us up the hierarchy of creatures,[51] Tesauro views the creative powers of metaphor as analogous to those of God Himself. A natural talent for witticisms, combined with artistic practice, makes a man so sublime "that he no longer appears to be a terrestrial Man; but a celestial God in his Art." God can make something out of nothing, but with the use of our "ingegno" we, too, "out of nonbeing, make being." Thanks to a limitless capacity to combine heterogenous parts into monstrous wholes through the wonder of metaphor, a human being "makes a Lion become a Man; & an Eagle a City. He grafts a Woman onto a Fish; & fabricates a Siren for the Symbol of Flatterer. He couples a Goat's torso onto the tail of a Serpent; & forms a Chimera as a Hieroglyph for Madness."[52]

Tesauro goes even further. Wonder is not only a pleasurable distraction, it is a necessity required to alleviate the inherent ennui of the human condition. Whereas an angel "speaks not with the Signs of concepts, but with the concepts themselves, so that one thing is both signifier & signified," our species alone, along with God, has the need and the capacity to signify things by means of other things. This, unfortunately, is not the place to explore Tesauro's fascinating and cosmic theory of semiosis any further. The point I want to make here is that "neither the Animals nor the Angels, but Men alone were given by Nature a certain nausea of everyday, although pleasant, things, unless utility is joined with variety, and variety with pleasure."[53]

Now, what most clearly distinguishes Peregrini and Gracián from Tesauro is not this need for variety and pleasure, which they all share to some degree; it is rather the specific cause of pleasure. For the moderates, like Gracián, "it is an act of understanding that expresses the correspondence that is found between objects"; and "to find correspondences between correlated terms is the foundation of all finesse": correspondence, harmony, correlations are the sublime sources for wonder.[54] And indeed the practice of *inventio* in rhetorical theory is founded on similitude. It is also, in part, a Renaissance aesthetic based on the microcosm/macrocosm analogy between the human body and the universe, especially popular in seventeenth-century Italy in the version of the "doctrine of signatures."

Tesauro is not opposed to this search for similarities, but to a large extent he continues the rhetorical tradition of searching for startling differences through resemblance to a third term. The most efficacious source of wonder for Tesauro is in many instances incompatibility, and not similitude. Describing the metaphor of opposition, he explains that "these are the kinds of witticisms that we may call in the Greek fashion THAUMA, that is, THE WONDROUS: which consists in a Representation of two Concepts, that are almost incompatible, & therefore superwondrous." In another statement regarding the desirable nature of incompatibility he describes "the MAKE-BELIEVE [*il fingimento*] in which by coupling incompatible terms, Enigmatic, Wondrous, & Ingenious Propositions will be born thereof. And like the basic Metaphor, it has the most incompatible terms, so that the Propositions will be more obscure, but more Wondrous, and fanciful [*capricciose*]."[55]

While Tesauro does indeed draw his taste for incompatibility and the definition of the oppositional (or antithetical) metaphor from the ancient *Rhetoric*, Aristotle's emphasis was always on a regulated, "balanced" pairing of terms. True, Aristotle does state that "an acute mind will perceive resemblances even in things far apart," and that "antithesis impresses the new idea more firmly"; but he also repeatedly warns that "persuasiveness . . . springs from appropriateness" and that "plainly the middle way suits best."[56] The question is not, of course, whether an orator should use appropriate figures of speech for persuasive ends, but just what constitutes that appropriateness. It is Tesauro's willingness to redefine those parameters of appropriate usage that constitutes both his continuity with the rhetorical tradition and his rupture with the ancient art of persuasion.

The association between wonder and persuasion is, as we have said, already encoded into Aristotle's *Rhetoric*, one of the founding texts for *seicento* formulations on wit. The rationale for giving "everyday speech an unfamiliar air" is to startle the hearer, since "people like what strikes them, and are struck by what is out of the way." By stirring up the emotions of the audience through the use of "strange words" among other things, "approval of course follows." The trick, however, is to neither lapse into the banality of "commonplaces," nor to stray so far from the common that your listeners become lost. Hence, defamiliarization through far-fetched metaphors is not only inappropriate, it is ineffective: "It is also good to use metaphorical words; but the metaphors must not be far-fetched, or they will be difficult to grasp, nor obvious, or they will have no effect."[57] This is where we can locate the radical and discontinuous character of Tesauro's *Cannocchiale:* for Tesauro there is no upward threshold to the persuasive effectiveness of startling metaphors; the more far-fetched or "pellegrino," the more persuasive effect they are thought to have on the hearer.

Tesauro departs from Peregrini and other predecessors in his enthusiastic embrace of extreme metaphors and excessive usage. Where Peregrini identified a defect, "seeking out rarity from what is furthest from the common," Tesauro sees a virtue: "And those who are most ingenious are the ones who are able to recognize & couple together the most far-fetched circumstances." The incompatible, the unharmonious, this is what generates marvel, and what else is such a composite of opposites but monstrosity? In oppositional metaphors, a negative and a positive combine to form "a monstrous composite; which because of its novelty generates wonder, & this, delight." Metaphors are monstrous, and monsters are metaphors: "Similarly wondrous are MONSTERS: Nature's witticisms."[58]

Actually, anything can be monstrous or not; the only limits to Tesauro's tremendous *combinatorio* are the limits of the human imagination. Kitchen utensils, for example, combined in the fashion of an Arcimboldo portrait to create "a capricious Grotesque witticism in human form that has a bucket for its bust; for a soldier's belt, a ring of casks, from which hang various utensils. For its neck, a box of salt; for its chin, a handle; for its teeth, a saw; for its beaked nose, a brush-hook; for eyes, two bowls; & likewise for the other parts."[59] It would seem that all the danger, all the allurement and threat of lapsing into the purely corporeal stupor brought on by the monstrously marvelous have been largely evacuated in Tesauro's view. Can it really be that only twenty-odd years since the appearance of Peregrini's treatise the monstrous has lost all its negative, antisocial valorizations?[60]

A closer reading of Tesauro's *Cannocchiale* shows that the ambivalence characterizing the new, rare, and unusual has seeped deeper into the signifying structure of tropes. In this dispersal of the grotesque it breaks down clear boundaries, not only between civil and uncivil behaviors and between man and beasts but also between things and words: they combine, fuse, and lose clarity. Ambivalence has become the operating semiotic principle. Tesauro's handling of the sirens, just one example out of dozens, shows how insidious and arcane the workings of the monstrous have become.

Fighting Fascination with Fascination

In the piazzas and from the pulpits, a battle was going on to wrest the erring faithful from the enthralling influence of popular entertainers: "Charlatans, histrions, tightrope walkers, prostitutes, pimps, *messieurs Alphonse*, thugs, and above all, venerable purse snatchers." In Naples, they made "the infamous Piazza del Castello . . . the habitual theater of their antics."[61] At least one priest was not to be intimidated and took to the piazza to fight the representatives of the devil on their own ground. Sir Gilbert Burnet, the Scottish historian who passed through Naples in 1685, narrates having seen "a Gesuit, Father Francesco de Geronimo da Grottaglie . . . making his way in the manner of a procession, [and] although he was well accompanied, he never gave up calling out to every-one who saw him and he exhorted them to follow him, and having arrived at a place where a charlatan distributed his drugs, and having taken his place, he jestingly entertained the people until the charlatan retired and left; then he quit, too, fearing that the gathering, having him as its sole entertainer, would be bored."[62]

Another contemporary witness gives an even better idea of the tre-mendous energy and charisma that De Geronimo habitually had to mus-ter on these occasions:

> The first enemies he declared an implacable war on were the his-trions, the charlatans, and the acrobats who . . . were accustomed to bringing onto the stage impudent, shameless women or young men, and with the obscenity of their comedies they are a great incentive to wrong-doing, especially for the impetuous and fiery youth. For which reason . . . with that freedom that his sacred ministry granted him, by jumping up onto the stage or standing nearby, starting to preach, he forced them to give up and yield the place to him, taking with him the huge audience. . . . He would draw out his sermons for as long as three or four continuous hours, without ever tiring and without ever being tedious to the audience, until, night having fallen, the charlatans no longer had a terrain in which to put their evil designs to work.[63]

Jumping on the stage, inspired by his sacred ministry, De Geronimo literally usurps the actors' place, leaving them no alternative but to yield their space in the public's attention to the orator and entertaining them for three or four hours at a time. What could he possibly have been saying, one wonders, until nightfall, never wearying and never boring his listeners, hour after hour, captivating a crowd of people that had originally come to the square, after all, to see some spicy theater?

Paolo Segneri, another Jesuit and perhaps the greatest sacred orator of the century,[64] devotes a sermon to the malevolent influence of these "istrioni": "Presently there are those who insinuate themselves onto our stages, some with trickery, with iniquity, with perfidy, and what is worse, they gain ground with sacrilegious spells, now reigning, now taking revenge, now conquering, and notwithstanding, one sees them arrive with fortunate outcomes to where they so maliciously aspired to be." But the scurrilous, worldly content of their performances was only one of the reasons Segneri and his colleagues staged such a violent reaction to "this pestilent pastime."[65]

The main reason Segneri urged his congregations to stay away from such shows was because he firmly believed that "the spectators of profane performances are all possessed by demons, not in their bodies, which would be a lesser evil, but in their spirits." "The actors are so many magicians of consciences" and "the listeners are all bewitched in their souls." In short, "what appear to be burlesque entertainments are truly enchantments produced by sorcerers' will." Strong words from a holy man who spent twenty-seven years, from 1665 to 1692, touring the Italian countryside barefoot "between toils, sacrifices, and hardships, often deliberately pursued for the purposes of mortification." The orator knew from firsthand experience the kind of spells and enchantments that actors could cast on the eager populace. His own mission in the dozens of villages he visited was not limited to oratory: "Upon his arrival, people would run 'processionally' from far away . . . because he was not only a famous preacher, he was also the judge who strove to reestablish peace, to placate disturbances and conflicts with diplomatic prudence."[66]

Segneri was a beloved patriarch, a prophet, a shaman. Like De Geronimo, who had to fight fire with fire, jumping up on stage to usurp the actors' place, but then who found himself forced to "outentertain" his rivals, Segneri was faced with a similarly equivocal role. The people flocked to him in processions looking to him for guidance and authority. If he was to maintain his unique place in their communities, he had to embrace a certain violent self-representation: "The virtue of Christians is a virtue of the cross, of contradiction, and of violence," he preaches; "the Kingdom of Heaven will not give itself except to he who obtains it with force and takes it by storm, weapon in hand. 'The violent shall make themselves the masters of it.' "[67]

What takes place in theatrical performances, and what the orators find themselves having to combat, is the same passion that Descartes and the moderate critics so deplored: excessive wonder. The terminology that Segneri uses has a slightly different cast, but the phenomenon is identical. Public spectacles were instituted, Segneri admits, "in order to indoctrinate the people into habits, approving good ones and condemning the wicked." Now if theatrical performances have the power to "impress the sentiments" that they promote, how exactly does this effect take place? The answer: through fascination, "a certain torpor of the senses [un certo intormentimento del senso]" which lulls consciences to sleep "as if drugged by opium [quasi adoppiate]" by "that bewitched lethargy [quel letargo incantato]" that results in "a kind of madness [un genere di pazzia]" and "a certain immobility [una certa immobilità]."[68]

Theatrical performances are a kind of opiate that induce lethargy, dulling the senses, silencing the reasoning powers of the conscience, bewitching the brain, and resulting in a certain madness and immobility. "Is it not an enchantment? Is it not a spell? Is it not a bewitchment? And then, who are those people that you allow to so twist you about? Who are they, I will say it like this, who are those magicians that have sickened your imaginations? . . . They are a mob of vile rabble."[69]

The actors may have been vile swindlers and cheap exhibitionists whose only motive was to siphon as much money as possible from their victims' pockets, but the method that their would-be rivals were forced to adopt in order to take their place differed not a whit from the actors'. Segneri, De Geronimo, and priests like them—Dominicans, Jesuits, and Capucins—had to outperform the performers, cast more potent spells, induce an even greater state of astonishment, provide an even more petrifying spectacle. ("Do not worry about having ample material to admire, to fear, to hope, to moan about, to exult, to stupefy you.")[70] They resorted to what were called *concetti predicabili* or preachable conceits.

Preachable Conceits

"Preachable conceits made up a spectacle," Benedetto Croce writes in one of his *Saggi sulla letteratura del seicento*.[71] "The academies praised the preacher, publishing collections of verse and prose; high society sought in the Lenten sermons a substitute for the diversions of carnival; rivalries between religious orders provoked enthusiastic partisanship among the public."[72] The focal point of these spectacles was a particular form of conceit adopted by orators for use in sacred oratory. The so-called *concetti predicabili* or *napolitani* were said to have arrived in Italy from Spain, by way of Naples; developed in the 1500s by Panigarola and other famous preachers, they provided a compromise solution and reaction to "the abuse of burlesque sermons with mimed performances and scurrilous scripts"[73] that had plagued churches at least since Dante's time.

Tesauro defines the practice as "this novel manner of teaching while delighting, & delighting while teaching, by means of these ingenious arguments, commonly called *Preachable Conceits*, which, with wondrous & new & metaphoric reflections on the Sacred Scriptures & on the Holy Fathers, lowers difficult doctrines to the capacities of Idiots & exalts the low & flat to the realm of the Learned: in the same way Manna pleases and nourishes the little people and the great, the nobles and the plebeians."[74] Their novelty consisted in basing the sermon on some more or less obscure middle term that did not come from the sacred scriptures, thus creating an artificial difficulty, then solving it through an interpretation of the ingenious metaphor. "I say that it is a Witty Conceit; that is, an Argument ingeniously proving a Proposition coming from Sacred material, & able to be persuasive to the People whose Middle term is based on Metaphor."[75]

To explain in detail: (1) take a theme suggested by God (e.g., that Christ was born when human evil had come to its ultimate point); (2) add a middle term or "Argomento ingenioso" based on one of the seven species of metaphor (e.g., the astrological circumstances present at the time of Christ's birth); (3) point out a difficulty that the middle term gives rise to ("Why would God have his most beloved son born during the most cruel, wintry time of the year?"); (4) offer an unraveling of the difficulty, showing that precisely what seemed most inexplicable and contradictory was in fact "a Divine Witticism, when it is well understood" (i.e., that God had in mind the future redemption and established the seasons in order that the winter night would symbolically correspond to the moral condition of the world); (5) next, the application, "which applies the opening argument to the Scriptural Passage, & the Step from the Scripture to the Theme," the part that most requires an acute wit in order to "couple together two things that appear to be far apart & moreover with marvelous clarity of terms & correspondences"; and finally, (6) cite some authority (in this case, Saint Gregory of Nyssa) in order to confirm the interpretation ("producing new stupefactions," Tesauro assures us) and to ensure that the explanation does not appear as a mere invention of the orator's "ingeniosità."[76]

The reasons these *concetti predicabili* were such crowd pleasers is "primarily because of the Metaphor which is naturally enjoyable. Next, because of the strange & unexpected application. Furthermore, because of the erudition of the speech. And finally because of the Coupling of the Authority of the Holy Father with your conceit." It is this unlikely coupling, much like the connection between moral precept, words, and figures in emblems, that is so effective in inducing wonder. The preacher should prolong step number six: "You can ponder it part by part, & inculcate these words in your Listeners with greater emphases." Tesauro stresses this point: delight ensues from "seeing coupled the words of the Saint with your own thought, like in Emblems, the Saying coupled with the Body, as I have already said."[77]

Tesauro is a virtuoso at preachable conceits, precisely because they require such a monstrously hybridized coupling of terms and material. After all, "the PREACHABLE WITTICISM is a Conceit that is nimbly suggested by Divine Wit, prettily revealed by human Wit, & refastened with the authority of some sacred Writer": it borrows from God, humans, and sacred writings, fusing all three into a marvelous whole. It "nimbly suggests," "prettily reveals," and then, "refastens," like some erotico-mystical striptease.[78]

"These Alluring Sirens"

All the more understandable, then, is that out of eight examples Tesauro uses to demonstrate the various species of *concetti predicabili,* two make use of monstrous creatures as their ingenious middle term: the metaphor of deception takes the foolish giants who built the tower to the moon, and the metaphor of equivocation uses the siren story. Let us take a closer look at how one would go about constructing a conceit based on equivocation.

The theme for today's sermon is "That the Pleasures of the World are Afflictions"[79]—a seemingly banal proposition suggested by Christ's words that the seeds that fall among thorns are choked by the cares and riches of life and their fruit does not mature (Luke 8:14). And who would believe, to use Saint Ambrose's question, "that delights, and pleasures are thorns, if Christ himself had not said it?" But reasonable arguments and church fathers, as convincing as they might be, are simply inadequate for providing a good show. What you want instead is to "demonstrate this Theme with a Preachable & witty conceit & make it new with the novelty of a metaphoric Reflection."

Start with a little display of erudition (and this is where the equivocation takes place): in Hebrew, the word for affliction is the same as the word for voluptuousness, or pleasure, that is, *tannim.* Then with some clever prestidigitation, you can show that the line in Psalm 43 (verse 19) "humiliasti nos in loco afflictionis" leads to the tautological statement "Afflict us in the place of our afflictions," a saying that is "inept, & useless to reasoning, & to your speech." Why would God have been so careless as to allow the language of Christ and the first fathers—Hebrew, that is—to lead to such an "Enunciatione nugatoria [idiotic utterance]"?

The unraveling of this difficulty lies in showing that, contrary to a first reading, "this is a wondrously ingenious & witty Proposition." Because *tannim*, an equivocal word, "with the same word signifies two different, actually contradictory things at the same time": to wit, *tannim* means both *affliction* and *siren*. And at this point, the preacher is advised to tell the story of the sirens:

> On the warm beach of the Tyrrhenian Sea, three most beautiful Nymphs would entertain, alluringly inviting Seafarers to turn their sails toward their shore, promising them a nest of Favors and Passions, a homeland of merriment, a port of tranquillity, a paradise of delights and pleasures: with triplicate ties of charming countenances, harmonious strings, and sweet voices, singing they enchanted, and enchanting they enchained, so that those unfortunates, who, having forgotten their affairs & beloved homeland toward which they had been sailing, gazing at no other Star than those beautiful eyes, nor following any other wind than the pleasing air of those melodious voices, by oar, by sail, by flight, with a quick pace, pushed their ships laden with rich wares and new hopes toward that shore. But lo! no sooner berthed, seeing around them nothing but a horrid desert, and bare reefs strewn with bare bones, and unburied corpses, here a hanging skull, there a rotting bust, elsewhere a body half alive, stricken by terror of Death more than Death itself, prey to those singing Beasts, they lost their ships, their wares, and finally their lives.[80]

One now applies the pagan fable to the sacred verse. "*Humiliasti nos in loco afflictionis*" thus becomes "Afflict us in the SIREN place, & with a novel & supercelestial Dialectic, unknown to the worldly Schools, joining together, actually identifying these two contrary notions as equivocal, & preaching one and the other as univocal, it forms this marvelous, but true Enunciation, the Sirens are Afflictions: that is, the Songs of the world are Groans; the Happinesses are Sadnesses; Sensual Pleasures are Torments: *Afflict us in the place of our Sensual Pleasures.*"

This would be enough to stun one's audience into a wondrous stupor of admiration at your ingeniousness, but Tesauro presses on. Not once, but several times, God calls pleasures by the name of sirens.[81] At this point you cite the various biblical passages and offer interpretations. In addition, the mysterious pagan poets figured the sirens as lascivious women because fraudulence or deception is characteristic of women. Whereas other writers like Homer figured the sirens as bird-women, combining fraud with gracefulness, and others like Horace figured them as fish-women, combining fraud and "momentary & fleeting lewdness." God depicted the sirens as "Womendragon, or Dragonwomen," combining extreme fraudulence with extreme cruelty. "The Sphinx of Sipilus, the Hippocentaur of Thessaly, and the Minotaur of Crete all yield to this Monster."

And here your point comes across most emphatically (this is step five, the application): "You may apply this Monstrous Siren of Sensual Pleasures to everything that with the same delights, & worldly inducements, consuming and devouring the bodily forces of Voluptuous Men, effeminates noble spirits, saddens hearts, enervates the forces, tramples the flesh, reduces to nothing the goods of fortune & of nature: so that the same sensuality harms the senses and is a torment to itself."

Once again, this would seem sufficient to stupefy one's congregation, but God's wit is bountiful. *Tannim* "signifies not only a deceptive Siren, & cruel Dragon, but also a barren Reef, & rocky crag for shipwrecks." A sublime equivocation, which shows that God wanted us to know that "Sensual Pleasures are the Reef where ships laden with precious acquisitions, that with prosperous wind sail toward the Heavenly Port, go to be wrecked."

The sermon has almost come to a close. What you must now do is give a stirring invective against voluptuousness and worldly pleasures, "showing that every pleasure, every joy, outside of God, is none other than affliction." And finish off with a warning for the future: "If the Sirens were called Companions of Proserpine by Ovid: happy Shipwrecks, charming Deaths, cruel Joys, by Martial: by Claudius, sweet Evils, flying Goods, pleasing Monsters, alluring Dangers, enjoyable Terrors: you may well call worldly Pleasures tasty Poisons, gloomy Joys, lugubrious Laughters, unhappy Happinesses, deceptive Sirens, Women Dragons."

"One should finally conclude," says Tesauro,

> that if the most prudent Ulysses, furrowing the Sea to speedily reach his wished-for Homeland . . . tightly bound himself to the Mast of the Ship & filled the ears of his Rowers with wax in order to not allow himself to be distracted from his straight path by the sweet voice of the cruel Sirens, likewise the wise Soul must tenaciously bind himself to the Mast of the Holy Cross & closing his ears to the enticing song of Sensual Pleasures, he must open them up to the divine Voice that calls him toward the Heavenly Port, where true pleasures are and where our thoughts are directed.[82]

The sixth, and final step: seal off "all this" with an apt saying from Saint Ambrose. A tour de force, a stunning display of erudition and wit, truly worthy of the definitive theorist of *concetti predicabili.*

Your audience is surely swooning by now, transported into a state of wonder that far exceeds the enchantment provoked by any mere physical acrobatics in the local square, pathetic tales of crossed lovers, or histrionic displays. An extraordinary sermon of this caliber cannot fail to induce an intense, pleasurable state of *admiratio* in its listeners. Certainly an orator at this level of skill corresponds to Tesauro's practitioner who combines natural talent with studied art, an orator so accomplished "that he no longer appears to be an earthly Man; but a heavenly God in his Art."[83]

But, if we think back a moment, this marvelous affect that turns the speaker into a god resembles very closely the dangerous and immoderate practice depicted by Peregrini. One of the insidious effects of unregulated metaphors, according to Peregrini, was that the user of witticisms, and not the witticism itself, became the object of wonder. As we have seen, the sacred orator is himself a kind of siren, beckoning passersby, like De Geronimo in the Piazza di Castelnuovo, to join him; gesticulating, doing tricks with his wit, entertaining with consummate art, in order to seduce his congregation into falling under his spell. "The goal that every orator must have in view . . . is to please the listener, to stimulate everyone, and enter into the good graces of the people . . . of everyone, in fact, with his presence, his nonchalance, with his voice and gestures," remarks sarcastically a disillusioned Spanish Jesuit.[84]

The fact is, in the battle that the church militant was waging against secularized pleasures, the popular entertainer and the priest became drawn into the same vortex: willingly or not, they exercised the same profession of fascination as magicians and sorcerers, and the secret desire of both was in fact to *become* the adored object of admiration, to take the place occupied by the rare and extraordinary, the new and unusual. What they wanted to usurp, what they desired to become, we might say, was the monster itself.

The other pernicious effect of the indecorous use of *acutezze*, according to Peregrini, was a leveling of the social hierarchy. This is exactly what took place in response to the use of *concetti predicabili:* they pleased everyone's tastes "equally the lesser and the great; nobles and plebeians."[85] That was what made them such a boon for the preacher who had to address an unevenly educated crowd: they functioned to level out discrepancies in the hearers, "lowering difficult doctrines to the capacities of Idiots & exalting the low & flat to the realm of the Learned."[86] The eventual outcome of such a leveling of hierarchy, such as what took place during festivals and carnival times, was unpredictable and often led to violence; certainly the church itself moved to put an end to the practice of such oratorical immodesty and excesses.[87]

Even more disturbing, though, is an important omission that Tesauro commits. In his exploration of the meanings and figurations of sirens (voluptuousness, bird-women, mermaids, dragon-ladies, lascivious women, reefs, pleasures, riches) there is one obvious and commonplace represen-tation that Tesauro left out, one that he himself makes use of elsewhere in his own treatise. On the first page of his 750-page opus, in the first paragraph in fact, Tesauro boasts that "there is none so savage and inhu-mane that, at the appearance of these alluring sirens, their horrid counte-nances do not become serene again with a pleasing visage."[88] The alluring sirens that he is referring to are *argutezze*.

Let us stop a moment and consider what has happened. *Argutezze* are the most civilizing force, they are the flower of civil conversation, a unique talent given to human beings, products of a faculty that makes our souls resonate with the divinity and transforms the most savage and inhuman dolt into a picture of pleasure: monstrous metaphors, bewitch-ing bi-membered dragon-women, also known as sirens, that lure sea-farers to the shores of putrefaction and death, enchanting passersby with their mellifluous yet deceptive music, stripping them of their acquisitions and wits and lives, are simultaneously agents of disenchantment, saving those who have lapsed into savagery and inhumaneness and restoring them to their human state. "This Monstrous Siren of Sensual Pleasure" (what does a powerfully ingenious *concetto predicabile* provoke in a man but "voluttà," albeit in the name of a good cause?) consumes men devoted to pleasure, devours their bodily strengths, effeminates their noble spir-its, saddens their hearts, enervates their powers, rots their flesh, reduces the riches of fortune and nature to nothing. And yet this same siren principle, the pleasure principle promised by the sophisticated intellectual pastime of preachable conceits, is meant to save sinners' souls from the "diabolical lethargy," the tendency to lapse into an immobile state of stupor that is proffered in public entertainments.[89]

As Ulysses tied himself to the mast of his ship, "& stopped up the ears of his Rowers with wax in order to not allow himself to be distracted from his straight path by the sweet voice of the cruel Sirens" so must we strap ourselves to the holy cross: "The virtue of Christians is a virtue of the cross, of contradiction and of violence used against the rebel appetites," as Segneri proclaimed so emphatically. Yet the violence we do to our rebel appetites is not a purely Christian notion. It stands at the origin of West-ern myths concerning humanization.

The Strain of Holding the I Together

In their *Dialectic of Enlightenment*, Max Horkheimer and Theodor Adorno show how the program of the Enlightenment to disenchant the world by dispelling all fears and superstition is inherently self-destructive: "Believing that without strict limitation to the verification of facts and probability theory, the cognitive spirit would prove all too susceptible to charlatanism and superstition, it makes a parched ground ready and avid for charlatanism and superstition."[90] This attitude was clearly at work in early modern Italy in the disputes between priests and entertainers, and as we have seen, charlatanism, for a while anyway, triumphed by recruiting its unwitting enemies into its own ranks. The same logic led to Peregrini's contradictory attitude toward *acutezze mirabili*: the more he decried them, the less able he was to detach himself from studying and describing them. His fascination was an embarrassment.

What Horkheimer and Adorno call the Enlightenment spirit (which they see as already at work in pre-Socratic cosmologies) bolsters itself through a belief in the mutually exclusive aims of self-destruction or self-preservation. Preservation is pursued at all costs: it is achieved primarily by reducing the world and nature to objects, to a kind of "standing reserve" for the progress of human technology: "Men have always had to choose between their subjection to nature or the subjection of nature to the Self."[91] Thus the self is reduced to a logical construct, in direct opposition to the alienated otherness of the world of experience. In this terrified, compulsive pursuit of knowledge that is use oriented and promotes preservation, pleasure can only be experienced as a threat. To experience pleasure is to expose the self to a powerful urge for disintegration, for sheer unconscious uselessness, conceptualized as a merging with the object, as a collapse of "distance from the thing itself, which the master achieved through the mastered."[92]

The seventeenth-century view of the ambivalent nature of wonder clearly illustrates this dichotomy. A "technological" pursuit of wonder, one that uses it as a means for the well-defined end of improving learning and memory—wonder in the service of acquiring useful scientific knowledge—was applauded. But the moment curiosity became an end in itself, as soon as it promised a purposeless pleasure—seen as a merging of the subject of consciousness with the object of admiration, and figured as a statuelike immobility—it was condemned. The fixation and lethargy that were said to ensue from astonishment are one way of describing what Horkheimer and Adorno call "the dread of losing the self":

> Men had to do fearful things to themselves before the self, the identical, purposive, and virile nature of man, was formed, and something of that recurs in every childhood. The strain of holding the I together adheres to the I in all stages; and the temptation to lose it has always been there with the blind determination to maintain it.

The narcotic intoxication which permits the atonement of deathlike sleep for the euphoria in which the self is suspended, is one of the oldest social arrangements which mediate between self-preservation and self-destruction—an attempt of the self to survive itself.[93]

Tesauro's description of the effect of monstrosity on the soul sounds eerily similar to Horkheimer and Adorno's: "WONDROUS are those who, leaving behind the ordinary laws of Nature, hold the soul equally suspended & in wonder."[94] Paolo Segneri actually saw this effect as an opiumlike state of lethargy and loss of judgment (a kind of "narcotic intoxication"), which he could only attribute to demonic influence. And yet he too was put in the paradoxical position of acting as a charismatic leader for small communities. His spoken words became a different kind of enchantment: they laid down the law.

If, as Horkheimer and Adorno claim, the products of civilization can be seen as a kind of barrier keeping the self from death and destruction, what civilization has to offer as a substitute is only a different kind of self-sacrifice and renunciation: obedience to the social order, and labor, as means of staving off "the promise of happiness which threatened civilization in every moment." Thus in the brilliant reading of the siren myth found in *Dialectic of Enlightenment* the sirens represent a call to unproductive knowledge. The siren song offers a pleasurable cognition that has not yet been codified as the "merely contemplative knowledge" that bourgeois ideology will call "art." In Homer's version of the myth, the sirens promise to explain "everything that ever happened on this so fruitful earth" (*Odyssey* 12.191). The lure of history that they possess—knowledge that has not yet been assigned the task of providing "the material of progress"—is thus also viewed as a lure to regression to a superseded era, "any reversion to which was to be feared as implying a reversion of the self that mere state of nature from which it had estranged itself with so huge an effort, and which therefore struck such terror into the self."[95]

Odysseus does the only thing he can do. By tying himself to the mast, he abandons himself to pleasure, to intoxication, while ensuring that he will not be able to stray from his productive path as a proprietor and as a leader of men. The price he pays for this cunning solution is impotence. Powerless to act on the knowledge he receives, he ensures that he will be divorced from practice, reduced to a mere spectator, a concert goer, an art critic. His men, whom he protects from the lure of regression by stopping up their ears, are protected by means of a different kind of enslavement: to their labor. They are indentured to the practical; they continue to row, while looking straight ahead.

Conclusion

The emblem with which I began this chapter makes seductive woman the danger for men: she seduces "by her attractive appearance, by words, and by brilliance of spirit." The harlot that Ulysses mocks and the scholar repels is also aligned, according to the misogynist paradigm traced out in the remarks of Peregrini's editor, with the libidinous beauties that incontinent eloquence offers up for prostitution, and with Stigliani's "pregnant women" whose inordinate appetite for "filth" provides the figure for gluttonous pursuit of literary stupor. Yet the siren is at the same time a figure, a metaphor—for metaphor itself. The siren provides the *form* with which to represent metaphor, but the *content* of that form is metaphor itself.

For the scholar, the threat is posed precisely in the practice of his trade and by the tool of his trade: by words, and especially by "candore," the quality of naturalness of style and brilliance of discourse. The more "natural" his style is, the more he resists the allurements of "unnatural" artifice; consequently the more sincerely or naturally artful he will be. A scholar may have nothing to do with bird-brained women of brilliant wit, but he remains nevertheless attracted by the promise she offers of irresponsible verbiage, excessive figuration, satiation verging on nausea, an orgy of tropological intercourse. "Lust carries with itself many monsters," but it also carries the possibility for the scholar's redemption in production—in the production of alluring metaphors.

Ulysses' trajectory is largely defined by the monsters he must avoid. It is precisely by resisting their attraction that Ulysses (and his fellow seekers of knowledge) are able to trace out their path and adhere to its limits. In the case of the sirens, his mission is to submit to their song while repudiating them at the same time. But without the challenge to repudiate them, there would be no voyage past their song, there would be no knowledge to be gained, and there would be no desire drawing the voyager to that spot in the first place.

Tesauro interprets the sirens as pleasures in general that call one off the path of virtue. His version emphasizes the bourgeois failings of goods lost, enterprises eschewed, and profits passed up: first the victims lose their ships, then their wares, and finally they lose their lives. The seafarers are not passing by in a daring quest for knowledge: they have set sail in ships "laden in rich wares" and at the time they were pulled off course they had been pointed toward home, most likely returning from a profitable trading expedition to the colonies. Their sin is to have forgotten their (business) affairs and to have exchanged "a peaceful port, a paradise of delights & pleasures" for their beloved fatherland. The message? Stay at your posts in the commercial machine; pleasure is bad for business.

The emphasis on the *patria* is fitting for such a loyal courtier as Tesauro, who spent his entire adult life in the service of the Principe Tommaso Carignano, dutifully recording the prince's military campaigns and acting as official historian for the regime.[96] Pleasure is bad for business, and it is equally bad for politics. The description of the beach with its souvenirs of horror subscribes to a fittingly materialist emphasis on the realities of the flesh: "a horrid desert, and bare reefs, strewn with bare bones, and unburied corpses, here a hanging skull, there a rotting bust, elsewhere a body half alive." The horror lies not in the actual danger of losing one's life but in the mental anguish of contemplating the future horrors represented by the butchered bodies on the reef. The victims' bodies are not what are mangled by the sirens, though; it is their *expectations* that are mangled by remaining unfulfilled: the merchants are "stricken by terror of Death more than Death itself." The sirens offer no real violence to their victims; they simply sap their hopes, thereby taking away their capacity to produce. Liberated from forced labor, the merchants succumb instead to forced leisure; deprived of their ships and wares, they have nothing more to live for.

This is where the true deception lies: not in the deceiving appearances of the singing beasts who enchain with their enchantments, but in the surprising, unexpected effects of the paradise that they offer by "consuming and devouring the bodily strengths of Voluptuous Men, effeminating noble spirits, saddening hearts, enervating forces, pounding the flesh, reducing to nothing the goods of fortune and nature." Although his is clearly a gendered tale, the message for Tesauro is not so much "Beware of seductive women" as "Beware of fixations. Beware of forgetting. Beware of straying from the everyday values of the polis and the marketplace."

While Alciati, Peregrini, and others remain dazzled by the seemingly paradoxical nature of wonder-inducing metaphors, Tesauro, by embracing monstrosity, is in certain ways better able to explore the dependence-abhorrence relation that we all maintain with it. By calling metaphors "these alluring sirens" that save us from brutishness, he also (rightly) exposes the sirens as *what we most need*. This basic ambivalence was played out on the social scene in the disputes between preachers and entertainers. The preachers needed the new, rare, and unusual to attract and seduce crowds. Similarly, the skillful speaker of witticisms made objects of his listeners in order to manipulate them, but they in turn made him, the speaker, into their object of admiration.

The truth is, there never really is a clear demarcation between subject and object. The intermediate trance state provoked by wonder and contained by so many rules of decorum and appropriate usage is really a permanent state of confusion that the self experiences in the face of an objectivized world: a world that it maintains qua object only at a terrible cost to itself. It renounces pleasure and blissful loss of self at the price of constant labor. This inevitable return to the self as logical construct is the same position of subjectivity that views sirens as deceptive monsters.

We fear monstrous metaphors because we think they will cause us to degenerate, to lose the solid boundaries of selfhood, to take the place of the admired object, and finally, to become monstrous ourselves. But really this is a false fear. The same false dichotomy has been played out, over and over again, in Homer's siren story, in Plato's rejection of the Sophists and poets, in the seventeenth-century debate over good and bad wonder, and in today's polemics over mass-media and video-game influence on impressionable minds.

The truth is, we need to believe in the danger of monstrosity in order to not allow ourselves to be distracted from our straight path, as Tesauro phrases it. Women, lovers, madmen, and the mob must be figured as monstrously other in order to allow the preacher (or the scholar) to feel more human. The secret desire to usurp that place of monstrosity, to become the admired object, is part of the game we play "of holding the I together" by imagining its disappearance. A sort of "fort-da" game we play with our civilized selves.

AFTERWORD

As we have seen, the monster is a concept that we need in order to tell ourselves what we are *not*. During the period of the scientific revolution in Italy the monster lost its ancient associations with the realm of the supernatural and became nothing more than an interesting object to be described and ordered in a newly secular vision of reality. The sacred monster did become extinct in the Western world, but monstrosity—the pervasive threat to humanity that looms on the thresholds of our animal past and our machine future—will never die out. One of the things that makes it hard to track the monster is that it is constantly changing its attributes. That is because we humans are constantly changing our attributes, too.

Monsters of today have retained an even more spectacular connection with the mechanical than ever before. Our imagination is pleased to think of itself as a kind of virtual reality. We extend our minds over the Internet, learning to live in an electronic projection of our senses. Our computers play chess with grand masters: we yearn to pit ourselves against our technological counterparts.

As we model our self-image on computer metaphors, so our monsters become more computeristic. Cybernetic creatures, half-human/ half-machine, fill the big and little screens of our popular entertainment venues. Our favorite contemporary monster of all—the extraterrestrial— always arrives in a spaceship born from a superior technology. More often than not, we envy their superior advancement. Perhaps the truth is that we all secretly yearn to be aliens.

But these monsters are all of our own creation and fashioned very much in our own image. That is why it is so important to be conscious of the metaphors we choose when we place such creatures under the rubric of "monster." Because whatever a monster may be, it is a thing to be feared. If we are destined as a species to merge with our technological creations and to become more machinelike, then it behooves us to love our machines. It behooves us to love our monsters as ourselves.

NOTES

Preface

1 I use the convenient shorthand expression of "early modern" to designate the period between approximately 1550 and 1750. Similarly, the "scientific revolution" is a generally accepted label that indicates the period in Europe spanning the lives of Copernicus (1473–1543) and Newton (1642–1727). For recent discussion on this topic, see Margaret J. Osler, ed., *Rethinking the Scientific Revolution* (Cambridge: Cambridge UP, 2000).

2 I am referring to some notes that Wittgenstein scribbled to himself in 1931 into his copy of Sir James Frazer's *The Golden Bough*: "The historical explanation, the explanation as a hypothesis of development is only one kind of summary of the data—their synopsis. It is equally possible to see the data in their relations to each other and to combine them together into a general schema without doing it in the form of a hypothesis about temporal development. . . . The concept of a clearly arranged cognitive schema is of fundamental importance for us. It signifies our form of representation, the way in which we interpret things. . . . This clearly arranged cognitive schema brings about understanding, which exists just in that we 'see the connections.' Thus the importance of the discovery of interconnecting links. A hypothetical connecting link, however, should do nothing in this case other than to call attention to the similarity, the connection, of the facts." (The only English-language source for these notes is in Tambiah, pp. 56–63.)

3 Maria Stampino and Fede Moino gave me invaluable help in rendering Kircher, Schott, and Bonanni's Latin into English.

4 Sahlins x.

One: The Origins of Monsters

1 Martin v–vii. Saint-Hilaire is the author of *Traité de Tératologie* (1832). He calls for a separate science based on the laws of corporeal organization: "Teratology can no longer be considered a division of abnormal pathology . . . consequently, it constitutes a particular branch, a distinct *science,* in the special sense that we have given to this word" (x). The discovery of the Hermapolis mummy is also alluded to and confirmed by Taruffi.

2 Martin 5.

3 Ibid. 9.

4 Ibid. 10. The best source for information on ancient practices and juridical formulations regarding monstrous births is Taruffi. See especially vol. 1, chap. 1, "Costumi e leggi."

5 See Girard for a modern, anthropological overview of the role of monsters in traditional

cultures. *Violence and the Sacred* also provides a unifying theory of monstrosity in general.

6   Céard, *La Nature et les prodiges*, provides an exhaustive treatment of this topic.

7   And not from *monstrare* to show, as Michel Foucault erroneously describes in *Madness and Civilization*. "Until the beginning of the nineteenth century . . . madmen remained monsters—that is, etymologically, beings or things to be shown" (70). Chapter 3, "The Insane," does, however, provide a brilliant analysis of how Western culture has linked "its perception of madness to the iconographic forms of the relation of man to beast" (77). I would agree with Foucault that in the late early modern period both madness and monstrosity underwent a process of interiorization: from being put on show in public spaces, to withdrawing into the deepest recesses of the so-called human heart and human condition. Much more on this in chapter 3 of this book.

8   See Lenormant for more on ancient divination practices.

9   See Ottavia Niccoli's chapter on the role of monsters in divination and propaganda in her *Prophecy and People in Renaissance Italy.*

10 Taruffi 1, 106. Also recounted in Varchi 99–100.

11 Taruffi 1, 111. Arnold Davidson explores the changing links between horror and the monstrous in *The Boundaries of Humanity* 36–67.

12 See Lovejoy, *The Great Chain of Being;* Yates, *Giordano Bruno and the Hermetic Tradition;* and Koyré, *Études d'histoire de la pensée scientifique.*

13 Browne, *Religio Medici* 26.

14 Ibid. 53.

15 Ibid. 24.

16 Quoted in Battisti 127. The original: "O voi che pel mondo gite errando vaghi / di veder maraviglie alte et stupende / venite qua, dove son faccie horrende, / elefanti, leaoni, orchi et draghi."

17 Browne, *Religio* 29.

18 Aristotle, *Physics* 2.199b.

19 Aristotle, *Physics* 4.767b–769b.

20 Cicero, *De Senectute, De Amicitia, De Divinatione* 2.33.

21 Ibid. 109–10.

22 Ibid. 118.

23 Ibid. 109–10.

24 Taruffi 1:7.

25 Niccoli argues that the apex and sudden collapse of popular prophecy in Italy took place between 1527 and 1530. She sees the prodigy and prophecy craze as rising out of both high and low culture sources: a renewed interest in classical sources (specifically *De divinatione*) but also the persistence of *divinitio vulgaris,* a popular science of signs relying on interpretation from elders of the community. Prophecy was certainly used as an instrument of propaganda, but in my view, the functionalist explanation is inadequate; prophecy was part of a "scienza unitaria" that connected nature with religion, and religion with politics. According to Niccoli, in Italy the phenomenon of popular prophecy died out with Charles V's coronation in 1530, whereas in France, Germany, and England monstra-mania took off in 1530 and continued until the mid-1600s.

26 Céard, "Tératologie et tératomancie" 3.

27 Céard, *La Nature et les prodiges* 60.

28 Ibid. 62.

29 Pliny §32.

30 Ibid. §8.

31 Ibid. §6.

32 See Robert Schilling's introduction to the 1977 edition, xvi–xvii, for a list of critical epithets.

33 Pliny §32. "These and similar varieties of the human race have been made by the ingenuity of Nature as toys for herself and marvels for us."

34 Ibid. §34.

35 Ibid. §188.

36 Friedman 24.

37 Augustine 662.

38 Ibid. 456.

39 Ibid. 449.

40 Quoted in Friedman 185.

41 Ibid.

42 Augustine 985.

43 Ibid. 982.

44 Ibid. 980.

45 Ibid. 982–83.

46 Ibid. 661–63.

47 Marino 133. This is just one of many animals that Marino uses to describe Luther.

48 See A. Schramm, *Luther und die Bibel* (Leipzig, 1923) fig. 3-26; quoted in Baltrusaitis *Réveils et Prodiges* 362 n.

49 Liceti, *De monstrorum* 258; Paré 41.

50 Baltrusaitis, *Réveils et Prodiges* 313.

51 Céard also briefly discusses this monster on pages 8 and 13 in his article "Tératologie et tératomancie." See also Niccoli's chapter on monsters.

Two: Monstrous Matter

1 Written by Giovanni Rucellai, Bernardo's son.

2 Machiavelli met there Zanobi Buondelmonte (to whom he dedicated *La Vita di Castruccio Castraccani*) and Cosimo Rucellai, Bernardo's nephew (to whom, along with Buondelmonte, he dedicated his *Discorsi*).

3 Such as Nerli, Gelli, Trissino, Guicciardini, "nonche molti altri uomini politici di lettere, di toga e di armi, fiorentini e forestieri" (Lucarelli 105). L. Passerini's *Memorie storiche* (Firenze, 1854) also provides information about the Rucellai gardens, although I was not able to consult it. Today what remains is a statue of Polyphemus (eight meters, forty centimeters tall) and two grottoes that housed philosophical academies (Battisti 416).

4 The passages cited from Malaspini's story (Novella 24, vol. 2, 1599) are taken from Berti 144–46, and from Battisti 134–35.

5 Battisti 134.

6 Battisti in fact refers to "quell'altro 'Bomarzo' fiorentino, che furono, ma ancor più nel seicento che nel cinquecento, gli Orti Rucellai di Firenze" (133). For more on the Italian garden see Basile, including a bibliographic overview. Other works of interest are Praz, *Il giardino dei sensi;* Ragionieri; and Coffin.

7 "Sopra la generazione de'Mostri, & se sono intesi dalla Natura, ò nò" in *Lezzioni* 85–132. See "Vita di M. Benedetto Varchi, scritta dall'abate Don Silvano Razzi" at beginning of *Lezzioni*. I am grateful to Judith Brown for clarifying several ambiguities regarding Varchi's description of Cosimo I.

8 Ibid. 97: "Diciamo esser verissimo che così negl'Animali come negli huomini nascono

parti monstruosi, i quali ò abbondano, ò mancano delle membra ordinarie, così esteriori, come interiori, ò l'hanno trasposte, ò offese."

9   Ibid. 97–98: "Ma che bisogna raccontare quello, che scrivono gl'altri? Non sene sono veduti molti, & anticamente, & ne' tempi nostri, non che in Italia (come fu quello di Ravenna) ma nel Dominio Fiorentino, & in Firenze medesima."

10 Ibid. 98: "Quanti sono in questo luogo, che si recordano d'haver veduto quel Mostro, che nacque dalla porta al Prato circa dodici anni sono."

11 Ibid. 98.

12 Compare the more businesslike later scientific prose description of the Royal Society's *Philosophical Transactions* 1:10. "An Account of a very odd Monstrous Calf . . . produced at Limmington in Hampshire, where a Butcher, having caused a Cow . . . to be covered . . . killed her when fat, and opening the Womb, which he found heavy to admiration, saw in it a Calf, which had begun to have hair, whose hinder Leggs had no Joints, and whose Tongue was, Cerberus-like, triple, to each side of his Mouth one, and one in the midst: Between the Fore-leggs & the Hinder-leggs was a great Stone, on which the Calf rid: The Sternum, or that part of the Breast, where the Ribs lye, was also perfect Stone; & the Stone, on which it rid, weighed twenty pounds and a half; the outside of the Stone was of Greenish colour, but some small parts being broken off, it appeared a perfect Freestone. The Stone, according to the Letter of Mr. David Thomas, who sent this Account to Mr. Boyle, is with Doctor Haughteyn of Salisbury, to whom he also referreth for further Information." Apart from the one mention of the poetic character of Cerberus, the rest of the description is highly quantified. Note also the pedigree of the information and the invitation to verify the account. It would be instructive to compare an Italian description of a monstrous birth of the same period. Unfortunately, the only accounts I have come across from Italian scientific academies do not treat deformed births.

13 Varchi 98: "quel Monstro . . . il quale fu ritratto egregiamente dallo eccellentissimo Bronzino."

14 I do not know if Bronzino's sketch has been preserved. His interest in abnormal anatomy is in any case evident in his portrait of "Morgante nano visto di schiena" (Florence, Museo Antropologico), reproduced in Berti, fig. 134.

15 The practice of employing artists for illustration as an adjunct to a scientific collection was later institutionalized by Francesco I, who kept a full-time artist (Jacopo Ligozzi) on staff in his Studiolo. In his "Discorso Naturale" (1572–73), Ulisse Aldrovandi pleaded for full-time professional artists as an essential instrument for scientific investigation. (Reprinted in Pattaro 175–232.)

16 Or was it older than that, having died a natural death? Varchi keeps the origin of the specimen shrouded in mystery.

17 Varchi 98: "Trovaronvisi due cuori, due fegati, & due polmoni, & finalmente ogni cosa doppia, come per due corpi, ma le canne, che si partivano da' cuori si congiugnevano circa alla fontanella della gola, & diventa[v]ano una: Dentro il corpo non era divisione alcuna ma le costole dell'uno s'appiccavano alle costole dell'altro infino alla forcella del petto, & da indi in giu servivano ciascuna alle sue schiene."

18 Ibid. 98: "Questi, & molt'altri Mostri simili, & diversi, come quello, che se vede nella loggia dello Spedale della Scala, crediamo noi filosoficamente, che siano stati, & che possono essere."

19 Augustine 622.

20 Much excellent work has been done over the past two decades on the Italian Renaissance garden. I base my very brief comments primarily on Marcello Fagiolo, *La città effimera*, and Ragioneri, both of which provide bibliographies. I make especial use of

two articles: Alessandro Rinaldi, "La costruzione di una cittadella del sapere: l'orto botanico di Firenze" (Fagiolo 193–202); and Fagiolo, "Il giardino come teatro del mondo e della memoria" (Fagiolo 125–41).

21 This and previous quotations from Riccio 65.

22 Fagiolo 133–35.

23 Varchi 99–100. "Quel Mostro, che nacque l'anno 1543, in Avignone . . . il quale . . . haveva la testa d'huomo dagl'orecchi in fuori, i quali insieme col collo, braccia, & mani erano di cane, & così il membro virile"; "tutte le membra canine erano coperte di pelo lungo, & nero, come era il cane, col quale confessò poi essersi ghiacciuta quella tal donna, che l'haveva partorito"; "visse tanto, che fu portato da Avignone à Marsilia al Cristianissimo Re Francesco, il quale l'ultimo giorno di Luglio fece abbruciare la donna, & il cane insieme."

24 Reproduced in Berti, figs. 134 and 135.

25 See Lugli, figs. 35 and 37.

26 The gender appellation is unfortunately all too accurate. There is not a single treatise on monsters written by a woman, from any epoch, of which I am aware. This fact in itself would make an interesting topic of speculation.

27 Alluded to and commented on by Aristotle in *De Caelo* 300b.26–301a.20; and *De. gen.* 1.722b.

28 I have strung together three separate fragments. From Barnes 180–81.

29 Lucretius 195–99.

30 For this distinction the best recent source is Tambiah. See also the classic studies by Malinowski; Bronowski; and Evans-Pritchard; for a stimulating theoretical accounting of the various schools of thought see Skorupsky.

31 Varchi 89: "È tanto necessaria questa ultima cagione finale, che tutti gl'effetti, che ne mancano, se bene hanno tutte e tre l'altre cagioni, Efficente, Materiale, & Formale, non percio si possono chiamare veramente naturali, non essendo intesi, cioè ordinati, & voluti dalla Natura, ma fortunevoli, & casuali, come prodotti temerariamente, & à caso, fuori della volontà, & intendimento del producente."

32 Ibid. 89: "essendo sozza, & rea cosa"; "errori, & peccati di chi gli fa." N.B. *Peccato* could also be translated as *defect* in this and other passages to follow.

33 Ibid. 89: "non potemo pensare, ne devemo, che siano ne intesi, ne voluti, ne da DIO, il quale non puo errare, ne dalla Natura, la quale mai non pecca."

34 Ibid. 108: "laquale senza dubbio opera per alcun fine, et nondimeno erra qualche volta, come un Gramatico, che non sempre scrive bene, ò paria correttamente, et un Medico qualche volta da una medicina, che non opera, ò opera il contrario dell'intendimento del Medico." Borrowed from Aristotle, *Physics* 2.199b, this illustration resurfaces in many subsequent writers.

35 Ibid. 109: "Alcuni altri dicono che i Mostri sono prodotti à significare, & annunziare le cose future, allegando gl'avvenimenti, che si leggono essere seguiti dopo cotali portenti, & prodigij in tutte le storie, & l'usanze de'Romani, i quali gli facevano, ò ardere, ò gittare in mare, ò portare in qualche Isola deserta, & abbandonata, per placare l'ira degli Dij, & fuggire il soprastante pericolo per ordine, & comandamento degli Aruspici."

36 Varchi 109: "Ma è da sapere innanzi procediamo piu oltra, che questo nome Natura . . . significa oltra la Natura universale, cio è DIO: la Natura particolare, & questa si divide in due, nella forma, che è agente, & nella materia, che è paziente. E dubbio dunque se i Mostri sono peccati della Natura, di qual Natura si debbe intendere ò dell'universale, ò delle particolare. Et se dalla particolare, di quale, della forma, ò della materia, ò tutte, & due insieme."

37 Aristotle *Physics* 2.192b–193b.

38 Varchi 113: "Potemo finalmente conchiudere, che i mostri essendo errori, & peccati, non sono intesi, ne dalla Natura universale, ne dalla particolare, le quali non possono errare, ma dalla fortuna, & dal caso."

39 Ibid. 113: "Non fa ciò per sua colpa, & difetto, ma impedita da altri." Aristotle also characterizes nature as beleaguered but he never commits himself so clearly to blaming the resistance of matter since that would violate the absolute inseparability of matter and form. Cf. Aristotle: "It is the aim, then, of Nature to measure the coming into being and the end of animals . . . but she does not bring this to pass accurately because matter cannot be easily brought under rule and because there are many principles which hinder generation and decay from being according to Nature, and often cause things to fall out contrary to Nature" (*De generatione animalium* 4.778a). In any case, nature operates according to general, regular laws: "In natural products the sequence is invariable, if there is no impediment" (*Physics* 2.199b).

40 Varchi 113: "In questo modo medesimo potemo dire della Natura, la quale non erra mai per se, perche se il seme è indisposto, & ella fa quello, che può deve essere scusata, & se il mestruo non è tanto, ò tale, quanto, & quale si recerca, che sia, che colpa v'ha la Natura?"

41 Ibid. 113: "non importa altro, se non mancamento d'ordine, & non conseguimento di fine, & in somma un mancare dal solito corso, e ordinario costume, la qual cosa benche non venga per colpa, ò difetto della Natura."

42 (Padova, 1616, 1634, 1668; Amsterdam 1665). I used the 1634 edition at the Marciana library in Venice.

43 See Paula Findlen's fascinating article "Jokes of Nature and Jokes of Knowledge: The Playfulness of Scientific Discourse in Early Modern Europe," in Findlen, *Possessing Nature*.

44 An idiosyncratic variation on the very popular Africa = heat = "semper aliquod novi fert Africa" theme ("Africa is always coming up with something new") is given by Liceti in *De monstrorum* 219–20: it is the burning heat of the air that drives all the animals, including man, into a venereal frenzy, leading to the disastrous results of interspecies coupling that have made Africa famous for the constant production of monstrous novelties.

45 Liceti, *De Monstrorum* 3–4.

46 Ibid. 60–61: "To begin with, one must not include amongst [the category of] 'uniform mutilated monsters of the human species' the dwarves that are commonly called Pygmies, because first off they appear frequently, and they come from small parents and do not frustrate nature, in any fashion, in her purpose, and secondly, they have neither deformities nor vices as their cause."

47 Ibid. 51–52: "One may easily conceive three sorts of monstrous productions. The first is supernatural, which God permits and procures, in order to punish men for the sins and crimes they have committed, as the Sacred Scripture teaches us happened to Nebuchadnezzar. The second are infranatural, when the devil creates an illusion for man's senses, like that seen in people transformed into others through Sorcery. And the third is purely natural, and comes from some defect or other kind of obstruction, located in the principles naturally destined to form the bodies of men. We will not undertake here to examine the miraculous Origin of Monsters, nor those that the devil procures through his illusions. We will only speak about the natural ones."

48 Ibid. bk. 2, chaps. 23 to 30; the following quotations are from pp. 125–27.

49 Aristotle, *Physics* 2.199a.

50 I will be quoting from the 1658 English translation. First published in short form in 1558, and an immediate best-seller undergoing at least five more Latin editions during the next decade and in translation in Italian (1560), French (1565), Dutch (1566), and, as Porta tells us in the preface, also in Spanish and Arabic. The full and expanded version, with twenty books, came out in 1589, and it went through at least twelve editions in Latin, four in Italian, seven in French, two in German, and two in English. The Price quotations are from pages vii and viii of his introduction.

51 Porta, *Natural Magick* 3.

52 Thorndike (6:156) notes that Porta's inclination toward the occult "seems to have brought him more than once to the verge of difficulties with ecclesiastical authorities, but he always escaped serious molestation or punishment by submission or by influence in high places. Thus the Accademia de' Secreti which met at his house and to which it is said no one was admitted who had not discovered something new in nature . . . was forbidden by the papal court because its members were given to illicit arts. Porta was denounced to the Inquisition for using demon aid but successfully defended himself against this charge, although it was repeated in 1581 by Bodin." (I doubt seriously that Porta resorted to demon aid; it is far more likely that, like Campanella, he perhaps indulged in conjuring friendly angels in order to interrogate them on the secrets of nature. Admittedly, it was not always an easy task to distinguish between good and bad demons!)

53 Ficino 85.

54 Ibid. 179.

55 This and the previous quotations, ibid. 87.

56 Ibid. 180.

57 Ibid., first part of quotation 87; second part 178.

58 Ibid. 180.

59 Ibid. 117, 106, 117.

60 Porta, *Natural Magick*; all quotations in this paragraph and the following from pp. 1–2.

61 See Zambelli for an excellent overview and discussion of the history of this debated distinction from medieval times to the late Renaissance.

62 Porta, *Natural Magick* 1–2.

63 Ibid. 26–27. This is Porta's paraphrase of the Lucretian story of mother earth: "The earth brought forth of its own accord, many living creatures of divers forms, the heat of the Sun enlivening those moistures that lay in the tumors of the earth, like fertile seeds in the belly of their mother; for heat and moisture being tempered together, causeth generation. So then, after the deluge, the earth being now moist, the Sun working upon it, divers kinds of creatures were brought forth, some like the former, and some of a new shape."

64 Ibid. 33.

65 Ibid. 48.

66 Ibid. 48–49.

67 Ibid. 50–51.

68 Quoted in Park and Daston 20–22. The authors give a full account of the brothers' triumphant European tour.

69 Reported by Lugli 139, who reproduces the engraving in fig. 126.

70 Ibid. 139.

71 Ibid. 111. Lugli, from whom I cull these examples, gives several instances of these "lusus scientiae" as she calls them, and provides illustrations as well (figs. 123, 89, 87).

72 Guazzo 26–27. The first edition was published at Milan in 1606; second edition, 1626.

Nothing is known of Guazzo's life except that he was a member of the Milanese Order Ambrosiani.

73 Liceti, *De monstrorum* 231–32: "A child may be born perfect: by the power of the demon and the illusion that he bestows, it could seem monstrous. . . . Next, the malicious demon could, through an incantation, either by taking away the infant, or by a feigned pregnancy, replace the infant being born with a monster. He could also provoke the birth of a monster by operating the play of active forces against those of passive forces, by bringing into the uterus of the woman some causes of monstrosity. The virtue of the generative semen is in this case so diminished that it is unable to constitute a fetus that resembles its parents. The demon could also make some part of the formative matter of the fetus be unsuited to take the form of that of its parents. He could make the nourishing virtue of the fetus unable, through his perturbation, to bring necessary nourishment to all the parts to give the members their exact shape, he could send to the body of the fetus an illness that would deform its image and give it that of an animal: he could imprint on the imagination of the parents such a vision that the fetus bourne of it would be monstrous; he could mysteriously introduce into the womb fertile semen from different animals, producing monsters; he could even add to the half-formed fetus the half-formed infant from the womb of an animal, in order to give to this human product a monstrous constitution."

74 George Luck (p. 163) explains that "in the early texts the distinction between daimon 'divine being' and theos 'god' was not always clear. By the later Hellenistic period, however, the distinction between theos 'god' and daimon 'evil spirit' had become fairly common. . . . Since Socrates himself explained his daimonion [a related word] as an inner voice that warned him whenever he was about to do something wrong, it could not be considered simply an evil power, at least not within the Platonist tradition, for for Plato states that 'every daimonion is something in between a god and a mortal' (*Symp.* 202E)."

Three: Monstrous Machines

1 Hans Ulrich Gumbrecht has suggested in conversation that perception of form in general, and especially perception of monstrous form, is precisely what makes a monster *not* terrifying, since form allows the possibility of "hetero-reference." (The term comes from Luhmann.) Self and other depend on the establishment of difference and distance. Only when something remains formless does it adhere mysteriously to and within the self, and this incapacity for heterogenous reference is the most truly horrifying phenomenon: the unlocatable and seemingly unregulatable presence of software, for example, or the anguish-provoking existence of mutating viruses in the body. Monsters, in this view, actually become a comforting apparition: being able to identify them, define them, and perhaps even dissect them, renders them harmless. For the monster as scapegoat see Girard, especially chapter 6, "From Mimetic Desire to the Monstrous Double," pp. 143–68.

2 This contention is at variance with other scholars who have examined the question. For other perspectives on this issue, see Céard, *La Nature et les prodiges*; Daston; Park and Daston; Wittkower.

3 Céard provides a sampling of canard or pamphlet titles taken from J.-P. Seguin's catalog *L'Information en France avant le périodique: 517 canards imprimés entre 1529 et 1631* (Paris: Maisonneuve et Larose, 1964). Of 517 listed by Seguin, 323 date from between

1600 and 1631; only 20 of these are about monsters, and only a few of these indicate that they derive from pamphlets originally published in Italy. Céard sees this genre as a popular version of the prodigy book, serving as intermediary between high and low culture consumers, reciprocally influencing and providing material for each other. Céard, *La Nature et les prodiges* 468–83.

4 For example, one might say that an extraordinarily talented individual like Michael Jordan is "a monster" in basketball: for some people, Jordan really is a "god"; there is something supernatural about his ability to fly through the air. For others, his monstrous qualities imply nothing more than a naturally endowed, hard-working, and disciplined athlete. Similarly, there is an aura of "magic" about Magic Johnson; and so on. . . .

5 Augustine 332.

6 Porta, *Natural Magick* 51.

7 Ibid. 53–54.

8 Ovid 243.

9 Porta, *Natural Magick* 54–55.

10 Campanella, *Del senso* 221: "But today this name is defiled by being applied only to the superstitious friends of demons, because people, not bothering to investigate things, have sought shortcuts through demons, for that which they are unable yet pretend to be able to do."

11 Ibid. 222: "The most studious Porta nevertheless made an effort to revive this science, but only in a historical sense, without giving an account of its causes; and Imperato's museum may be partly a base for rediscovering it." On the importance of Imperato's museum for Campanella: "Dava la più grande importanza al celebre Museo che F. Imperato teneva in sua casa . . . e che egli avea dovuto visitare, come del resto la visitavano tutte le persone non ignoranti che venivano a Napoli" (Bruers's note, 222). For more on Imperato's museum, including a reproduction of its interior, see Olmi and Ferrante Imperato, *Historia naturale*, 1599.

12 Campanella, *Del senso* 223.

13 Ibid. 240.

14 Ibid. 242.

15 Ibid. 272.

16 Ibid. 272.

17 Ibid. 304.

18 Ibid. 307.

19 Ibid. 305.

20 Ibid. 306.

21 Ibid. 54.

22 A highly hypocritical conceit, given the homosexual relations that Campanella is known to have maintained at least with his close friend Fra Pietro Ponzio. A spy, sent to transcribe Campanella's nocturnal conversations while the prisoner was being observed for insanity, records the following exchange:
Fra Tomase: "Bona sera, bona sera."
Fra Pietro: "O cor mio, come stai? che fai? Sta de buono animo, che domano vieni lo Nunzio qua e saperimo qualche cosa."
Fra Tomase: "O fra Pietro, perché non opri qualche modo, e dormimo insiemi, e godemo."
Fra Pietro: "Volesse Dio, e dovesse dare dieci docati alli carcereri; e a te, cor mio, te vorria dare vinte basate [baci] per ora."
The modern editor of these Dialoghi notturni notes that "homosexual relations were

common in the monasteries of the time, and no less so in the prisons." Firpo, *Il supplizio* 177–83.

23 Campanella, *Città del Sole* 53–54.

24 Ibid. 58.

25 Ibid. 116.

26 Ibid. 114.

27 Ibid. 120.

28 Campanella, *Del senso delle cose* 309. "E chi sa argomentare nelle plante e nelli animali farà nascere a suo modo la prole e magnificherà o abbasserà la generazione umana con quest'arti; e se la donna desiderosa di qualche cosa, molto si affligge di voglia, perchè il portato suo è seco una cosa, s'infà pur della stessa voglia, e s'è tenero ancora, la viene ad esprimere nel lavoro, e, dicono, in quella parte dove la donna desiosa si tocca. . . . Or chi sa mettere desiderio di grande o picciola cosa nella femina pregna, può far nascere il parto con figure e colori di quella cosa; e se desidera virtù, nobilità e sapienza communica al parto il medesimo; e così lo spirito del fanciullo con quell'affetto lavora. . . . Questa affezione porta il fanciullo fuori e diventa tale qual si è procurato dal mago."

29 "Plus ça change, plus c'est la même chose." For an update on reproductive technologies and the woman's role in them see Janice G. Raymond, *Women as Wombs: Reproductive Technologies and the Battle over Women's Freedom* (San Francisco: Harper, 1993).

30 Campanella, *Del senso delle cose* 310–12: "Possibile è al mago, come fa nascere le piante dove li pare, far nascere anco animali; ma solamente quelli che di putrefazione si fanno, come sorci, pidocchi, mosche e serpi, ma non cavalli, elefanti e uomini. Si dice che Archita fè una colomba che volava con l'altre, e in Germania aquila a tempo di Ferdinando Imperatore si è vista; ma questo si è fatto con contrapesi, come gli orologi, chè, volgendosi alcune ruote attorniate di fila, alzano e abbassano l'ale; ma non credo io che avessero spirito e carne. Così le statue di Dedalo con contrapesi e argento vivo si movevano. Pur si dice che Alberto fece parlare una testa d'uomo; ma è favola: non può, si non nell'utero, farsi animale tanto perfetto. . . . Avicenna contende che possan nascere e così Epicuro e altri, come nel Nilo allagato superbi animali si fanno dall'umido terreno. Delli uomini selvaggi si può pensare, e di quei del mondo nuovo, che non si sa come andaraono là. . . . Nondimeno non so istoria sicura e penso essere impossibile. . . . Quest'arte di far animali ancor non s'è vista, chè già saria rimasta tra gli uomini. Pariavano le statue con li demonii inchiusi, e non bisogna di ciò dubitare, perchè spirito umano in metallo non può capire. Pigliar in lambicco li spiriti d'animali dentro uccisi, non è possibile per la gran sottilezza, nè rifonderli come spirito di calamita. Nascere da sè sì gran statue belle, mi par troppo. Li satiri saranno uomini selvaggi, come si vede spesso esser stati presi e aver imparato civiltà e generato uomini, e io ne vidi in Roma; e da fanciulli smarrite restano selvaggi."

31 Ibid. 356. See Baltrusaitis, *Le miroir*, for a complete history of the "errors of mirrors" theme, begun by Seneca, presented in theorem form by Euclid, and continued by Alhazen in a treatise of that name. Notes Baltrusaitis, p. 293: "Les erreurs d'un miroir sphérique concave (Alhazen, VI, 38; Vitellion, VIII, 25, 26, 40, 41, 52) sont presque toutes de la tératologie. Les agrandissements qui s'y produisent sont illustrés par l'oeil de Cyclope, la face de Bacchus, un doigt pareil à un bras ou à une pyramide d'Égypte. Les 'trompeuses grandeurs' servaient selon Sénèque aux jeux d'une luxure dépravée. Porta [1561] et Le Loyer [1588] en ont repris les descriptions osées."

32 Campanella, *Del senso* 361: "Set your head below that point, and you shall behold a huge Face like a monstrous Bacchus, and your finger as great as your arm. . . . Seneca reports that Hostius . . . so order[ed] his Glasses, that when he was abused by Sodomy,

he might see all the motions of the Sodomite behind him, and delight himself with a false representation of his privy parts that shewed so great."

33 Ibid. 286.

34 Campanella, *Città del sole* 106.

35 The edition I consulted is from 1676. It consists of four volumes on optics (538 pp.), acoustics (432 pp.), mathematics (815 pp.), and physics (670 pp.).

36 The 1667 Herbipoli edition that I consulted has 1,389 pages. Thorndike (7:591) mentions four editions coming out between 1662 and 1697—a remarkable success, considering their enormous size.

37 This is the list, taken from the table of contents of *Magia universalis*, that Thorndike put together in vol. 7, p. 598.

38 Schott, *Physica curiosa* 1:12: "Verùm ut nihil fuit unquam tam sanctum, laudabile, & incorruptum, quod Daemonis fraude, & hominum improbissimorum malitiâ non fuerit vitiatum atque corruptum; ita reconditarum quoque rerum scientia ab Adamo posteris suis tradita, & Magiae nomine cohonestata, nefariis superstitionibus, & execrandis dogmatibus Daemone Magistro contaminata fuit, sicque Magia primaeva atque legitima cum nomine coepit it idium verti apud bonos, brevique adeo diffundi pestis per orbem, ut nulla unquam gens extirerit, quae illâ non fuerit contaminata. Quâ ex corruptione nata est divisio Magiae in licitam & illicitam."

39 Ibid. 1:18.

40 Ibid. 1:22.

41 Ibid. 1:22–24.

42 Ibid. 1:27.

43 Ibid. 1:18.

44 Schott, *Magia universalis* 1:211. There are plenty of these spell recipes to choose from, all with the same characteristics.

45 Kircher, *Ars Magna Lucis et Umbrae* 906. He boasts that "imò non in tot se figura ipsemet Proteus transformat, quin in plura semper speculum hoc phantasticum te sit transformaturum." Proteus is often invoked in this context. Lorenzo Legati (p. 213) refers to a spherical convex mirror that hangs from the vault of the Cospi museum and "unisce in un punto la figura di che vi mira, e rappresenta in compendio tutto il Museo [unites at one point the figure of whoever looks into it, and it represents in a compendium the whole Museum]" as "Proteo bizzarrissimo [most bizzare Proteus]."

46 Baltrusaitis, *Le miroir* 19.

47 Godwin 13.

48 John Webster, *Metallographia* (London, 1671) 30; cited in Thorndike 7:568.

49 Gabriel Clauder, *De tinctura universali* (Altenburg, 1678), 58–60; cited in Thorndike, 7:568.

50 Schupbach 174.

51 Bonanni 3. "Tria insuper hic longo conclavi adiecta sunt cubicula, quorum primum machinas continet, & experimenta, quae Artes mechanica, & hydraulica subministrarunt. In secundo codices manuscripti dispositi sunt antiqui, & diversarum linguarum volumina Siriacae, Hebraeae, Graecae atque Sinensis; libri etiam permulti qui Statuas, Numismata, Gemma ostendunt, imò in eaodum visuntur numismata Imperatorum, Summorum Pontificum & Virorum illustrium, quibus Historia illustratur. In tertio denique Automata diversa." This description comes from Bonanni's later catalog of the reorganized and restored collection, but I believe, looking carefully at the illustration of the 1678 catalog, that it is also accurate for the original disposition of the collection.

52 Schupbach 174, from Bonanni's "Proemium" 1.

53 Bonanni 309: "Aliqua hujus generis [Authomata Instrumenta] in hoc Musaeo asserva-
bantur, quae per neglectum obliata vel moraci temporis injuria perierunt. Restituta
sunt tamen, & plura sunt addita caeteris non inferiora."

54 Kircher, *Ars magna Lucis* 901: "Retulit haud ita pridem non nemo ex familiaribus,
librum se vidisse Ioanni Trithemio adscriptum, in quo Author transformationem hominum in quodcunque animal promitteret; neminem tamen rationem assertionis capere
potuisse: unde id nisi artibus diabolicis fieri minimè posse plerique autumaverint.
Quicquid sit de observata exhibitae metamorphoseos ratione, hic disputare nolo: tantù
dico multa esse in rerum natura . . . quae tamen à solis naturae arcanorum conscijs facilè
in effectum deduci queant. Novi multa Trithemio impia affingi, quae tamen tantùm
abest, ut suspecta sint; ut nihil potius naturae magis consentaneum esse videatur, qué-
madmodum in Arte nostra Combinatoria, volente Deo clarissimè demonstrabimus. Ego
arbitror promissum Trithemij duplici via posse comperi, vel arte catoptrica, vel ar-
caniore quadam rerum applicatione, qua homo se in aliquod animal conversum putet."

55 Baltrusaitis, *Le miroir* 34: "L'homme y voyait d'abord le disque solaire, symbole de la
puissance cosmique, les bêtes y défilaient ensuite en alternant avec son visage propre
qui paraissait changer continuellement. Nous sommes en pleine métempsychose."

56 Kircher, *Ars Magna Lucis* 901.

57 Ibid. 905: "Si hoc speculum secundum latitudinem introspexeris, sine fronte primò te
ipsum, deinde asininis auribus conspicuum reddet: naribus, & ore nihil deformius esse
potest; nam ita sinuosè producentur, praesetim si rictum dentium monstraveris, ut
saxosum maris littus videri possit: subinde bicipitem, & tricipitem dabit: verbo, monstruosa apparitionis varietas vix verbis explicari potest."

58 Ibid. 905: "Si caprinam faciem exhibere desideres, speculum in duos umbones ex plano
aliquantulum undulato protuberabit, spectabisque sub Satyri se forma, turpem, cor-
nutum, rugosum, & oris hiatu ridiculum; rubicundam quoque, & incensam ebri faciem,
si rubro folio sublinieris, exhibebit. Cervi caput aspicies, si speculum in umbones
ramosos efformaveris. Uno verbo, nullum monstrum tam turpe est, sub cuius forma te
in speculo simili industria adornato non respicias."

59 Ibid. 906: "Scias igitur res quasdam naturales esse, quae mox ubi per os assumsptae
fuerint, imperium in phantasiam exercentes, hominem In id transmutent, ad quod vel
maximè inclinaverit. . . . Sunt & aliae res, quae per cibum sumptae, in Feles, Canes,
Lupos dicta ratione transmutent, atque; transforment. Verùm cùm haec extra artis
nostrem limites constituta sint, & talia quoque sint, ut ob multa mala, quae inde
emergere possent, ea propalari nec debeant, nec possint, in ijsdem tantùm summam
illam, & admirabilem naturae maiestatem venerantes; ea summo, perpetuoque silentio
consecramus."

60 Guazzo 50–51.

61 Baltrusaitis, *Le miroir* 38.

62 As witnessed by another experimentalist priest, Frère F. Jean François Nicéron, "Pari-
sien de l'Ordre des Minimes." Nicéron's book 3, *La Perspective curieuse ou magie artificièle des effets merveilleux* (1683), begins with the following discussion and definition
of "La Catoptrique et des miroirs": "La Catoptrique ou science des miroirs nous a fait
veoir des productions si admirables, ou plustost des effets si prodigieux, qu'entre ceux,
qui l'ont cogneue & practiquee, il s'en est trouvé, que par une vaine & ridicule ostenta-
tion, ou quelquesfois pour abuser les plus simples, se sont efforcez de passer pour
devins, sorciers ou enchanteurs, qui avoient le pouvoir, par l'entremise des mauvais
esprits, de faire veoir tout ce qu'ils vouloient, fut-il passé, ou à venir. Et de faict on en a
veu des effects si estranges, qu'à ceux, qui n'en sçavoient pas la cause, ny les rasons, &

n'avoient iamais rien veu de semblable, il devoient passer pour surnaturels, ou bien estre reputez pour de pures illusions ou prestiges de magie diabolique."

63 Baltrusaitis, *Le miroir* 38.

64 Scarabelli and Terzago 34.

65 Battisti 220. Chapter 8 is devoted to "Per una iconologia degli automi." Battisti also includes the essential bibliography on automatons. The citation he reproduces is from a paper presented to the Società Filosofica Romana in 1958 by Valentino Braitenberg. The place of automatons in the history of philosophy alone is a dense topic, one which I will engage in only insofar as it converges with the figure of the monster as alter ego to the human. A few examples that stand out: René Descartes in *Traité de l'homme;* and a long passage in a letter of March 1638 (*Collected Works* 2:39–41) in which Descartes asks us to consider the judgment of a man who had grown up all his life without seeing an animal, "& où, s'estant fort adonné à l'estude des Mechaniques, il auroit fabriqué ou aidé à fabriquer plusieurs automates . . . qui imitoient, autant qu'il estoit possible, toutes les autres actions des animaux." Julien Offray De La Mettrie in his *L'Homme-machine* (1749), by pushing the Cartesian mechanistic model of the functioning of the human body to a reductio ad absurdum, turns the human being instead into a simulacrum of the machine.

A modern philosophical treatment of Descartes's use of the automaton as human-looking nonhuman other posited as a test case for the limits of skepticism is provided by Cavell 398–420. Cavell's discussion focuses on the question "What is the nature of the worry . . . that there may at any human place be things that one cannot tell from human beings?" (416).

And finally, many of Philip K. Dick's short stories dramatize in fictional form the very anxiety that Cavell alludes to. See *Robots, Androids, and Mechanical Oddities* and of course *Do Androids Dream of Electric Sheep?,* which provided the scenario for the film *Blade Runner.* (I am grateful to Kenneth Gross for suggesting that I look at the Cavell text.)

66 Battisti 247.

67 Ibid. 226.

68 Lucentini 156.

69 Battisti 226.

70 Ibid. 226.

71 Lucentini 158.

72 Mechanical invention has always been associated with inducing sacred wonder in the West: "Religion in ancient Crete and Greece and Rome supplied priests who were masters at applying new mechanical principles to their own practical purposes. These practical purposes were of two kinds: sometimes involving performance of ritual with an absolute minimum of priestly effort; sometimes an attempt to impress worshipers and patrons with divine signs and seeming miracles in the temples" (Brumbaugh 9).

73 Bonanni 309: "Authomata Instrumenta illa sunt, quae ita ab Arte sunt composita, ut videantur sponte moveri."

74 Ibid. 309: "Non minori oblectatione oculos, & auditum allicit machinula alia supra mensam collocata: Simiam illa refert militari Timanistriae habitu indutam. Haec cum denticulus ferreus lateri aptatus leviter tangitur, illico in iram exardescere videtur, totisque viribus tympanum apte dispositis ictibus pulsat, quibus milites ad pugnam solent excitari. Caput interim circumagit, oculos quaqua versum movet, & rictus aperiendo dentes ad morsus paratos ostendit."

75 Ibid. 310. "Dum eas circumspicit spectator, patet repente porta, & post illam specus horrenda in qua informe monstrum catenis devinctum, quod dum serpentem collo

advolutum removere conatur, terribilem reboatum tunc edit, cum Draco alius rictum illi admovet, ut mordeat."

76 Ibid. 310. "Objectum sane pueris met[u]culosis terribile, Viris verò ludicrum, sed Philosopho Mechanico non indignum ob artificium, quo motus, & voces perficiuntur."

77 Ibid. 310. "Ab hoc non longe distat loculamentum, ex quo deforme Sagae caput spectatoribus illudit exerta ex ore lingua, ignitis prominentibus oculis, & voce terribili emissa, qua metu perculsis ligneus catellus suis latratibus respondet adeo apte emissis, ut non rarò veri canes ab eo excitati veros latratus eisdem conjunxerint."

78 Scarabelli and Terzago 156.

79 "As a concept which expressed a pattern of activity transcending the strict confines of museum itself, the idea of musaeum was an apt metaphor for the encyclopaedic tendencies of the age. . . . [M]usaeum was an epistemological structure which encompassed a variety of ideas, images and institutions that were central to late Renaissance culture" (Findlen, "The Museum" 59).

80 See Pomian, especially chapter 1: "Entre l'invisible et le visible: la collection" 15–60.

81 Bartoli, "Della ricreazione del Savio," 13, 441–42.

82 Scarabelli and Terzago 156–66. Mummies were very popular in the seventeenth century, as much for the precious bitumen they contained (believed to have miraculous medical properties) as for their exotic aura of ancient mystery.

83 Ibid. 5. "Dall'esperienza può qualunque curioso appagarsi de'più mirabili effetti la mente, là dove in distanza allo Specchio proportionata distendendo col braccio la mano, subito gli pare di vedere, che un braccio, ed una mano dallo Specchio se n'eschino, e con tale differenza s'incontrino, che la destra distesa rassembra, & è creduta la sinistra. . . . Nell'avicinarsi di chiunque, benche di pigmea statura, vedesi con tutto ciò altretanto ingrandito, quanto, che di tenue corporatura notabilmente ingrossato. Se poi nella parte convessa v'è chi si affaccia, scorgesi tanto in vicina, quanto in rimota distanza senza danno di se medemo impicciolito, e ricevendo i raggi Solari, non con unirli, ma con allargarli li riflette."

84 Ibid. 6. "Specchio di vetro di configuratione quadrangolare ed oblonga nell'esterna superficie piano, e nell'interna rigato in quadro, dal che rompendosi l'imagine dell'oggetto in ciascheduna area delle cinque, che dimostra, non senza monstruosa confusione rappresentasi ò tutta, ò in parte, la faccia di che vi si rimira, onde con la latina espositione *plurimas tibi facies, multiplicatos oculos, geminatos nasos repraesentat*, potrebbe avverarsi di lui quel detto *Sol d'apparenze abondo.*"

85 Ibid. 6.

86 Ibid. 7. "Ingrandisce dalla concava l'imagine rappresentandola ad ogni poco rivolgersi dello Specchio in diversi modi al naturale sì, ma sproporzionatamente monstruosa, il che dalla convessa parte succede, mentre impicciolita l'imagine dell'oggetto in istravaganti diversità si tramuta."

87 Ibid. 7: "Da questo Specchio renderebbe non men motivo di riso Democrito, che Heraclito di pianto per insieme ridere, e piangere la mostruosa inconstanza dell'humana conditione, di cui direbbe Ovidio Nihil est, toto quod per stet in orbe / Cuncta fluunt, omnisque; vagans formatur imago." Translation from Ovid, *Metamorphoses* 15.371.

88 Ibid. 9. "Specchio, in cui mirandosi uno, benche di candidissima carnaggione, si vede riflettere la propria imagine annerita à guisa di un Ethiope."

89 Ibid. 35. "Piedestallo, nella cui parte superiore rinchiuso si mira un capo d'horribil Mostro; col semplice tocco di un grilletto, ecco immantinente aprirsi una porta della quale uscendo sì mostruoso Capo con terribil rimbombo di voce, che da se stesso tramanda, riempie chi l'ode di spavento; Da due cannoncini, che da entrambi gli orecchi

gli pendono, trattone un filo, escono furiosamente due Vipere, che frà mille ritorte divincolandosi non poco terrore a'riguardanti arrecano; e questi mentre del lor timore esser vana la cagione contemplando, con giocondo riso ripigliano dell'animo la già conturbata quiete, quando dal subito disserarsi di una picciola finestra, che sopra al detto capo si osserva, in nuovo sconvolgimento si abbattono, facendosi all'improviso vedere un più mostruoso Capo, che snodando dalla sboccatura de'labri una lingua, e contravolgendo frà le spaventose sue ciglia l'occhio fiammeggiante, e movendo gli orecchi, che d'Asino porta, sono di nuovo invitati ò allo spavento dall'apparenze, ò dal riso dal giocoso inganno."

90 Ibid. 36–37: "Oltre all'ordinato sconcerto de'suoi movimenti rende in un punto medemo stupidi, e l'occhio, e l'orecchio di chi v'assiste per qualche durata di tempo," "con non poca ammiratione de' Spettatori."

91 Ibid. 34, 13, 9, 10, 10.

92 E.g., the passage on the uses of mirrors, 208; the description of the catoptric theater that turns a soldier into an army, 209.

93 Ibid. 183. "Poiche in universale sono produzzioni ingegnose d'una Cagione fertile ne' suoi effetti, & avveduta non meno della Natura medesima, di cui elle è Vicaria, & Imitatrice così diligente, che non l'agguaglia solo, mà sovente la sorpassa nella bellezza, e perfezzione delle sue faciture, e non di rado corregge gli errori, benche involontari, di quella. Ond'è, che hà potuto gareggiar con essa di pregio, e pretenderne qualche fiata la preminenza, vantando non pochi motivi di maggior nobiltà."

94 Ibid. 27. "Onde la vita di questo Mostro, quantunque breve, fù un lungo argomento, che vivamente provava ne gli animali essere maggiore la necessità del Core, che del Cielabro."

95 Ibid. 5. "seguendo la Madre, che spirò l'anima nell'atto del partoririo, quasi che non si potesse dare all luce un Parto così stupendo, senza comperarne la nascita con la morte."

96 Ibid. 6. "parimente Nana, d'egual simmetria delle membra benissimo proporzionate alla de lei statura, che non giunge a trenta oncie Romane, con tutto ch'elle sia d'età d'anni LV, vivendo col fratello al servigio del Sig. Marchese."

97 Legati adds that their parents were very grateful to have the two monsters off their hands since they were of modest means. Are we to detect some sense of discomfort in the fact that these two human beings were paid to be perpetual specimens all their lives? It is difficult to tell. Legati does seem anxious to stress how well they were treated. He goes out of his way to mention that Cospi "had them educated and kept them in his service their whole lives" (7).

98 Ibid. 30. "come già fece il Petrarca quelle della sua Gatta, che anche a' nostri tempi si conservano."

99 See Jay Tribby's wonderful article "Cooking Clio and Cleo: Eloquence and Experiment" for a look at the culture of "civile conversatione."

100 Ibid. 209. "che sovente copia da un Gigante un Pigmeo, come ne'Convessi; da un Pigmeo un Gigante, come ne'Concavi semicircolari; ò da bellissima faccia un mostruosissimo volto, e talvolta un Caos di confusissime linee, tutt'altro rappresentanti, che fattezze humane, come in alcune de' Concavi Cilindrici."

101 Ibid. 209. "Chi non dirà presti l'ali a gli sguardi humani, acciò volino ove senza esse giunger non potevano? e, quasi dissi, a veder l'invisibile?"

102 Ibid. 215. "s'è osservato, quanto dispregievole per la deformità del ceffo, e per l'orridezza di tutto il corpo, altretanto mirabile per lo straordinario numero de gli occhi."

103 Ibid. 215. "E questi non fuori di se, ma in se stessa, fatta ricettacolo, ed insieme Teatro delle più maravigliose Opere della Natura, che in essa si scoprono."

104   Ibid. 215, 210.

105   Ibid. 457. "S. Fulgentio riferisce dell'origine dell'idolatria presso gli Egizii, volendo, che un tal Sirofane, uomo ricco, spinto dal soverchio amore verso un figliuolo unico estinto, per allegerirsi dal dolore fattasi fare una statua del defonto, questa venisse da'servi adulatori con corone di fiori prima adornata, e con offerte d'incensi poscia venerata: e che finalmente ad essa, come ad asilo ciascuno recorrendo, fosse per divina riconosciuta."

106   Aristotle, *Physics* 1.192b.

107   Ibid. 192b.

108   Ibid. 194a.

109   And one that would also describe Campanella's plan for state-instituted eugenics for the production of an ideal race: it is a subtle distinction based on ends and means that harkens to Schott's way of telling licit from illicit magics: If the end is good, the magic is good, etc.

110   Ibid. 192b.

111   Aristotle, *On the Soul* 412b.

112   Kircher, *Musurgia Universalis* 306. "ore oculisque mobilibus totoque corporis situ vitam spirans . . . statua perpetuò garriet, iam voces humanas proferendo, iam voces animalium, iam ridere & cachinnari; nunc cantare, subinde flere & eiulare."

113   Ibid. 306–7: "Varij varia de hoc mirifico machinamento commenti sunt; Qui se-cretiorum philosophantium dogmata sequuntur, id omninò fieri posse putant. Nam Albertum Magnum caput hominis, quod articulatas voces perfecte pronunciaret, ad-mirabili solertia confecisse asserunt; Aegyptios quoque, varias statuas confecisse articulatè quidlibet pronunciantes, in Oedipo nostro Aegyptiaco multis modis os-tendimus; Nonnulli tamen id veluti naturae legibus contrarium repudiantes, simile machinamentum fieri posse minimè sibi persuadere possunt; Alberti autem, Aegyp-tiorumque machinamenta vel supposititia & falsaria machinamenta fuisse, vel daemo-num ope architectata, eo ferè modo, quo daemonem olim per oracula & statuas arti-culata voce responsa dedisse legimus, asserunt. Multi tamen putant statuam fieri posse eo ingenio architectatam, ut voces aliquas articulatè pronunciet. . . . Quicquid sit de famoso illo Alberti Magni capite, coeterisque Aegyptiorum machinamentis dis-putare nolumus, sed ipso facto istiusmodi prodigium maximè in opus deduci posse asserimus, & ne verborum tantum ampullis id promittere videamur; hoc loco statuam fabricari docebimus."

114   Ibid. 306. "Oculorum observabunt motum, labiorum lingueque mirabuntur agilita-tem, totius corporis vitam spirantis compagem cum stupore intuebuntur; Verum quo artificio aut condita sit statua, aut quam abditam motus sui machinationem habeat, nemo deprehendere poterit."

115   Homer 385–86, lines 417–20 and 375–76. Lucentini notes that "l'elemento magico è deliberatamente soppresso in favore di quello tecnologico" (152).

116   Cited in Luck 22.

117   Ibid. 98.

118   Campanella, *De Homine* 159, 161.

119   Eusebius, *The Preparation of the Gospels* 4.1.6–9, cited in Luck 123.

120   Brumbaugh 5: "Could a mechanism move itself, and so imitate the distinctive be-havior of living things? And if it could, did this mean that the secret of the creation of a 'soul' had been penetrated by the mechanic? . . . The designers of automata seem to have become progressively more ambitious, and their work more admired. Finally, by about the second century B.C. they aspired to nothing less than duplicating the most

creative forms of human behavior with their self-propelled series of mechanical components. . . . This project of creating beings with souls mechanically is a strand in the
history of Western 'gadgetry' which played a part in forming new ideas of the soul,
the self, and the difference between nature alive and nature mechanical."

121   Lynn White also demonstrates that "necessity is not the mother of invention, since all
necessities are common to mankind living in similar natural environments. A necessity becomes historically operative only when it is felt to be a necessity, and after prior
technological development makes possible a new solution." For this reason, the "technological thrust of the medieval West does not yield easily to explanation" (175).
White locates the West's unique technological explosion in the Middle Ages to prescriptive beliefs embedded in Christianity: "It was axiomatic that man was serving God
by serving himself in the technological mastery of nature. . . . technological aggression,
rather than reverent coexistence, is now man's posture toward nature" (199).

122   Winner 20.

123   Ibid. 20.

124   Aristotle, *Politics* 1253b.

125   Ibid. 1254a.

126   Ibid. 1253b.

127   Ibid. 1254b.

128   Augustine, *City of God* 335.

129   Winner 21.

130   Ibid. 21, quoting George Kateb. For an update on this theme, see Ray Kurzweil, *The
Age of Spiritual Machines: When Computers Exceed Human Intelligence* (New York:
Viking, 1999).

131   Ibid. 34.

132   Ibid. 37, quoting Marx.

133   Ibid. 36–37, quoting Marx.

134   Ibid. 34.

135   Quoted in Winner, 42.

Four: Medicine and the Mechanized Body

1   "Now the good is always beautiful, and the beautiful never disproportionate; accordingly a living creature that is to possess these qualities must be well-proportioned"
(*Timeus* 87B–C).

2   "I presume we shall say a man was just in the same way in which a city is" (*Republic*
4.441d).

3   "Just as in the city there were three existing kinds that composed its structure, the
money-makers, the helpers, the counselors, so also in the soul does there exist a third
kind, this principle of high spirit, which is the helper of reason by nature unless it is
corrupted by evil nurture" (*Republic* 4.440e–441).

4   *Republic* 4.444d–e. Anthony Parel (101–3) also points to the theory of the humors,
deriving from Galen and Hippocrates, as the foundation for pre-modern political
theory and provides a succinct historical overview tracing a line from Plato, Aristotle,
Thomas Aquinas, and Nicholas of Cusa to Savonarola, Machiavelli, and Guiccardini.

5   Foucault, *Madness and Civilization* 21.

6   Ibid. 27.

7   Ibid. 70.

8   Ibid. 77.

9   Porta *Della Celeste Fisonomia* 181. "Volgarmente è cosa trita tra i Filosofi, che il mostro nel corpo è mostro nell'anima, qual mostro nell'anima, che cosa può aspettare dopo di se, che debba avvenirgli, se non mali, & infortunij."

10 Thorndike also reports interest in physiognomy in Spain, Germany, and France as paralleling that in Italy. See volume 8, chapter 35, "Physiognomy" (pp. 449–75) for an overview of the publishing history on this topic.

11 First printed in Greek in 1545 as part of Aelianus's *Varia historia* and subsequently translated into Latin and Italian by Carlo Montecuccoli in 1612, followed by editions in 1623 and 1626 and four other Italian editions up to 1652.

12 Porta, *Fisonomia* Proemio. See also Francesco Stelluti's dedication to Cardinal Barberino in his synoptic edition of Porta's *Fisonomia*: "La Fisonomia dunque è quella che n'insegna a conoscere la natura di detti segni; scienza utilissima, e necessaria, e spezialmente à Principi, e Signori grandi, havendo eglino più degli altri huomini privati di molti servi, e ministri bisogno, li quali se di prave qualità saranno, tanto maggiore sarà il numero de'loro occulti nimici: e per ciò devono procurare d'eleggerli buoni, e fedeli, e spogliati d'ogn'interesse."

13 Porta, *Fisonomia* Proemio. "Sarà utile dunque questa scienza non solo a conoscere chi habbia ingegno a che arte, o scienza inchinata sia laonde, si dovria nelle bene ordinate Città ordinar huomini di gran prudenza, che da quelle semplici fattezze de fanciullini scopressero qual arte, o scienze propria se gli convenisse."

14 Ibid. Proemio. "I Spartani havenano per usanza non darsi libertà al padre, & alla madre del figliuolo, se lo dovessero nodrire, ma era bisogno portarlo ad un tribunal deputato dove sedendo in consiglio quelli ch'erano maggiori nel giuditio, se vedevano nel figliuolo dalle proportionate fattezze del corpo, & del molto segni di robustezza alta a poter difender la loro Republica, overo d'ingegno, e di consiglio di saperia governare, lo restituivano ale lor madri a nodrire, ma se altramente, lo portavano nel monte Taigeto, e dirupandolo dalla cima, prima che al basso fusse gionto, moriva in mille parti dissipato, giudicano indegno di vita, chi non fusse atto giovar a se, ne alla sua patria."

15 Ghirardelli, *Deca prima*, Proemio. "Se questa gran machina Mondiale ha sempre mosso i curiosi intelletti a discorrere, e le saggie lingue a ragionare; molto maggiormente questo picciol mondo dell'Huomo, in cui pare, che il tutto ridotto sia (quasi in un compendio di perfettioni, e prerogative) dalla potente mano di Dio, che per questo fu termine dell'opere sue: molto maggiormente dico ha ciò operato, non solo destando sottilissime questioni sovra la propria sua creatione, e la parte di se più nobile, che, Anima, viene nominata; ma formando altresì mille profittevoli, & argute consideratoni sopra la parte materiale, e corporea, e de'suoi accidenti, che quasi misteriosi caratteri sono, per li quali delle interne passioni dell'animo sogliono venire i saggi in cognitione."

16 Thorndike 7:53.

17 Stelluti 6–7.

18 Ibid. 8. "Si vede per esperienze, che l'anima nostra sempre patisce al patir del corpo, & il corpo vien travagliato, & afflitto ogn'hora, che l'anima prova notabili passioni; e che medicandosi il corpo, l'anima inferma si sana, come avviene nella pazzia infirmità dell'intelletto: . . . onde si vede esser frà loro una grandissima confederatione, e fratellanza, e corrispondenza."

19 Ibid. 10.

20 Ibid. 18. Giambattista Vico, I would imagine, is an example of a person with a hot, dry brain, due to his choleric temperament. See the following chapter for a physiognomic analysis of Vico's features and physique.

21 Ibid. 23.

22 Ibid. 27.

23 Ibid. 97.

24 Ibid. 98.

25 Ibid. 99.

26 Ibid. 151. "Come gli Huomini Selvaggi, lasciati i lor ferini costumi, possino divenir mansueti, e piacevoli."

27 All consulted at the Bancroft Library at Berkeley. My thanks to the generous staff for providing these reproductions.

28 Porta, *Fisonomia*, 1652 ed., 318.

29 Ibid. 1616 ed., 175.

30 Ibid. 1668 ed., 175.

31 Ibid. 175. "Ciascun sà che la natura dell'unghie nasce dalla superfluità, e progressi della natura nascono dal calore, la strettezza dell'unghie è seguitata da ignoranza, e rozzezza, perche in loro la dolcezza è assai debole, ne può far in loro molto progresso, perche si possono dilatare, e far ampie superfluità, però coloro, che sono di sì poco calore, sono ignoranti, e stolidi, perche ogni freddezza, aporta seco stupore, & ingegno rozzo, e però coloro, che strette unghie haveranno, saranno ignoranti, e poco ingegnosi. . . . Plauto parlando d'un cuoco ladro. *E tu pur cerchi di trovar un cuoco. Che non habbia unghie dell'Aquila ò di Nibbio.*"

32 Ibid. 1616 ed., 164. "Quelli che si distorceno, e s'inchinano, sono adulatori, e così fa il Cane mentre adula, e blandisce. . . . Quelli che si abbassano, e rompono il corpo con tal modo, sono adulatori. . . . Io gli rassomigliarei alle Simie, e direi che sono buffoni, e maligni, che sempre si muovono, e distorceno il corpo, & imitano l'humane attioni."

33 Ibid. 1616 ed., 194.

34 F. Imperato, 10. "Quelle cose dalla natura prodotte, che han similitudine, & dimostran la forma, & effigie di molte cose naturali."

35 Ibid. 10.

36 Ibid. 83

37 Ibid. 33–35. "Il Signor Dio creò li nostri primi Padri, da quali hebbe origine la humana generatione; & in progresso di tempo il mondo abbundò di gente, che hebbero, & hanno la loro ordinaria statura, & altezza; & pur à tempi di Noè, vi furono li Giganti . . . li quali hebbero la loro altezza molto sproportionata dalla detta ordinaria; perche non può la natura adoprarsi nella generatione de Pygmei? li quali essendo di altezza di trè palmi in circa, non s'allontana molto dalla ordinaria altezza delle genti; nè si può riputare molto sproportionata, conforme à quella delli Giganti; oltra che vedemo la Natura compiacersi più della breve statura, che della grande; il che vien confirmato, con il veder la statura, & altezza humana sempre andar diminuendo, & à poco à poco gli huomini son ridotti alla statura più bassa, ch'era nelli primi secoli; del che Homero non cessava di ramaricarsi; e gionto con l'altezza son mancati gli anni, e la salute; e se ben li Giganti sono stati in più, e diverse parti del mondo; pur delli Pygmei si può dir l'istesso."

38 Ibid. 36–37. "Pochi anni sono mi venne referito da un molto Reverendo Monaco Augustiniano di nation Lapponese, il quale nel viaggio, che fè per Fiandra, gionse per tempesta di mare in un luogo, vicino all'Isola de Islan . . . ; ma per violenza del vento fù trasportato più oltra, ove ritrovò un sicuro ricetto, e quivi dimorò alcuni giorni, e volendo riconoscere detto luogo, se era habitato; e come era il suo nome, e se era Isola, ò continente; lo ritrovò deserto, & inhabitato; per il che entrò alquanto entro terra; & questo lo spinse principalmente l'haver ritrovate alcune piccole capanne, composte di legni, & altre di ossa di pesci del geno Cetaceo, intessute con bello artificio di certi gionchi più grossi dell'ordinario; ma caminando ritrovò un cespuglio, dentro al quale

vidde due, che in prima vista parevan pastori, vestiti di pelle di estrema bianchezza; e perche con velocità fuggirono, non li fù concesso di posser bene osservare le vesti, ne anco la loro effigie, & altezza; ma li parve la lor statura esser molto breve."

39 These quotations are contained in the title.

40 Ghirardelli, *Cefalogia fisonomica* 384. "Huomini, o voi, che dalle Donne il piede / Torcer mai non volete, immersi ogn'hora, / Anzi sommersi in quell'Egeo, che fuora / Lusinga, empia Sirena, e dentro fie[l]e. / Fuggite, incauti; ivi è d'Amor la sede; / Non già di quell'Amor, ch'in Ciel dimora, / Ma ben di quel, che sempre l'Alme accora. / Di fallacie ripien, voto di fede. / Fuggite, e più dell'altre, oime, fuggite / Quelle, che'l naso han rilevato, e corto; / Che son vane, lascive, e troppo ardite. / Trovan nel gloriarsi il lor diporto, / Sono a gl'inganni, & al furore unite, / Dicono mal d'altrui, ma sempre a torto."

41 Ibid. 624. "La Donna è poco differente dall'altra, cioè tutte le Donne non sono molto differenti ne gli affetti, niente ella havrà di moderato in se stessa; una Donna, o ama smisuratamente, overo odia mortalmente, non ha mezzo, nata solo a gl'inganni, ed alle frodi."

42 Ibid. 624. "La Donna è un mostro di natura, è Huomo imperfetto, come piacque determinare a tanti Savij, potremo arguire da tutte le sue parti."

43 Ibid. 623. "Non è dubbio veruno, che in picciol vaso, poco liquore si contiene; come anche dell'opposto a tutti è a bastanza palese . . . & essendo il Capo picciolo angusto spatio del Cervello, quindi è, che poco ancora diciamo esser il Cervello, come materia contenuta, che dal suo continente la quantità riceve. Da che formasi il giuditio, che lo spirito animale, e la forza di operar ciò, che da esso dipende, nella strettezza sua si soffoca, e quasi estinta ne rimane, e di qui nasce la debolezza, imperoche (come dice l'Ingegniero) tutte le nostre operationi sono serve, che essequiscono la cognitione della mente, & ove per indebito ministerio de gli spiriti, l'anima non può discernere la verità delle cose, nè si certifica dello stato loro, ella teme nel fatto, nè s'apprende a risolutione alcuna, che non dubiti fosse stato meglio far' il contrario (passione propria delle Donne)."

44 Ibid. 624. "Sono curiose de' libri profani, e massime di quei, che trattano d'amore, e ciò deriva dall'appetito libidinoso, che le predomina. Bramano le novità; ogn'hora vorrebbono mutar veste; inclinano all'otio, a i diletti, & a i piaceri, alle musiche, a i suoni, a i giuochi, non alle facende feminili, e della Casa; sono larghe nel donare. . . . Godono di esser servite, e di usar spesso delle cerimonie."

45 Ibid. 624. "Poiche sono bellissime parlatrici, in guisa tale, che terrebbono a bada qualunque più ritirato, e divoto spirito, & ammollirebbono con le sole parole ogni più indurato cuore, dicendo talhora concetti sì vaghi, e pizzicanti, che un Signor nostro amico ci affermava, che sovente esso havea commercio con una di queste tali, solo perche era invaghito del suo gratioso favellare, anzi che alcuna fiata notava in carta qualche bel detto, per servirsene a suo proposito nelle lettere; ma è ben vero, ch'egli stesso giurava di haver comprato le parole di quella Donna, col proprio sangue, poiche lusingato da esse, qual'Augellino dal simulato sibilo dell'Ucellatore, caddè nella rete d'amore, e la caduta fù tale, che non potè districarsene senza rimaner stroppiato, e nella vita, e nella robba, e nella fama. Il che sia a tutti essempio per viver lungi quanto mai si può da queste false Sirene; che a noi tanto basta haver detto per fine di questi Discorsi."

46 Porta, *Fisonomia*, 1652 ed., Proemio. "Habbiamo letto appresso gl'antichi Socrate Filosofo haver usato lo specchio per la buona institution di costumi, ilche fù ancora accettato da Seneca, che l'huomo possa specchiar se stesso, perche conoscendo le nostre imperfettioni ricorriamo al consiglio, & all'emenda. Conciosia che mirandosi alcuno in uno specchio, e vedendosi ben formato dalla natura, procuri per l'avenire, che non

imbratti la bellezza del corpo con la brutezza de'costumi, così veggendo il corpo brutto, procuri con ogni suo sforzo, e diligenza che con le virtù, medichi e risarcisca i brutti segni del corpo. Perche queste cattive inclinationi del corpo, come l'orgoglio, l'invidia, la superbia, infermità dell'anima agevolmente si possono curare."

47 Ibid., Proemio. "Habbiamo letto appresso gl'antichi Socrate Filosofo haver usato lo specchio per la buona institution de costumi, ilche fù ancora accettato da Seneca, che l'huomo possa specchiar se stesso, perche conoscendo le nostre imperfettioni ricorriamo al consiglio, & all'emenda. Conciosia che mirandosi alcuno in uno specchio, e vedendosi ben formato dalla natura, procuri per l'avenire, che non imbratti la bellezza del corpo con la brutezza de'costumi, così veggendo il corpo brutto, procuri con ogni suo sforzo, e diligenza che con le virtù, medichi e risarcisca i brutti segni del corpo. Perche queste cattive inclinationi del corpo, come l'orgoglio, l'invidia, la superbia, infermità dell'anima agevolmente si possono curare."

48 Ibid. 8. "Nell'infermità del corpo l'anima solamente si muta, che non può usar il suo ufficio, e quell'huomo, che la patisce, non è l'istesso, ma divien altro."

49 Ibid. 8. "Per molta, e straordinaria passione molti huomini sono divenuti peggio che bestie. . . . Nella Cinantropia, e Licantropia infermitadi, il molto si mutano in cane, gli occhi divengono infocati, con grigni minacciosi, e naso acuto; escono di notte, vanno intorno i sepolchri, ne si sente altro da loro, che latrare, ringhire, & altre cose."

50 Ibid. 7–8.

51 Bound together with the 1652 edition of Porta, *Fisonomia* 45. "L'anima, la quale dà l'essere al corpo, & in lui si vive, ella stessa è quella, che il dispone, non pur come sua casa, & suo domicilio, ma l'addatta al suo dosso, come fa il sarto alla persona altrui un giuppone, overo una calza; ond'egli è verisimile, che in ogni sua parte egli le risponda. Et sarebbe imperfetto l'animale, quando tra l'anima, e il corpo suo non si trovasse una vera corrispondenza; perche la materia non ubbiderebbe alla forma."

52 Porta, *Fisonomia,* 1652 ed., Proemio. "Scrisse Platone, e da lui Aristotele, che la Natura dà il corpo proportionato all'attioni dell'animo, conciosiacosa, che ogni stromento, che si fà per altra cosa, e tutte le parti del corpo per altra cosa son fatte, e questa per cui cagione si fa alcuna cosa, è una attione, onde chiaramente ne segue, che tutto il corpo è stato dalla natura creato per alcuna eccellentissima attione."

53 Ibid. 1.8. "Dunque la disposition del corpo risponde alle potenze, e virtù dell'anima, anzi l'anima & il corpo con tanta corrispondenza s'amano frà loro, che l'uno è cagion del gaudio, e del dolor dell'altra."

54 Porta, *Della celeste fisionomia* 106. "Onde giudico, che è determinato nella filosofia, che quel che manca d'alcun membro principale, è mostruoso, e mal fortunato, & che gli manca ancora qualche cosa dell'antivedere, & prudenze: & a quei, che sono mal costumati, & poco prudenti sempre accascano cosa nocive."

55 Ibid. 1.555. "Sappiate che l'huomo è stato locato dal sommo Dio, ne'confini dell'intelligenze supreme, e de' bruti; perche per l'intelletto sa'ssomiglia a quelli, per il senso a questi."

56 Ibid. 1.556.

57 Ibid. 1.556. "Noi diremo di quella virtù, ch'è tesoro d'ogni virtù, che è sopra di noi, e la chiamono heroica, e Divina. Questa si oppone a quel vitio, che habbiamo detto di sopra. Questa virtù per avanzar la nostra humana conditione, fà l'huomo, nel quale alberga quasi simile all'Angelo, overo alle divine intelligenze; perche quell'huomo ch'è pieno di tante virtù, par che avanzi la nostra humanità . . . si chiama Heroe, overo mezzo Dio."

58 For the formulation of these questions and reference to Joseph Needham below see Dini 135.

59 For the influence of Descartes on Neapolitan medical thought see Torrini. For the influence of Harvey in Italy and Naples see Schmitt and Webster, with an extensive bibliography; Chauvois; and Pagel and Poynter.

60 Harvey 2.

61 Ibid. 105.

62 Ibid. 107.

63 Ibid. 84–85.

64 See Thorndike 7:515 for a discussion of this issue.

65 Harvey 21–22. "Those two motions, the one of the ears, the other of the ventricles, are so done in a continued motion, as it were keeping a certain harmony, and number. . . . Nor is this otherwise done, than when in Engines, one wheel moving another, they seem all to move together; and in the lock of a piece, by the drawing of the spring, the flint falls, strikes the steel, fires the powder, enters the touch-hole, discharges, the balls flie out, pierces the mark, and all these motions by reason of the swiftnesse of them, appear in the twinkling of an eye."

66 Harvey 81–82.

67 Ibid. 46.

68 Ibid. 82.

69 Cited in Pagel, "The Reaction to Aristotle," 500.

70 Ibid. 503.

71 Harvey 84. "And hence perchance the reason may be drawn, why in those that with grief, love, cares, and the like are possessed, a consumption or continuation happens, or cacochymie, or abundance of crudities, which cause all diseases and kill men. For every passion of the mind which troubles mens spirits, either with grief, joy, hope, or anxiety, and gets accesse to the heart, there makes it to change from its naturall constitution, by distemperature, pulsation, and the rest, that infecting all the nourishment, and weakning the strength, it ought not at all to seem wonderfull if it afterwards beget divers sorts of incurable diseases, in the members, and in the body, seeing the whole body in that case is afflicted by the corruption of the nourishment, and defect of the native warmth."

72 Descartes, *Oeuvres et lettres* 734–35.

73 Descartes, *Oeuvres de Descartes* 1:236.

74 Ibid. 132–33.

75 Descartes, *Oeuvres et lettres* 209–25.

76 Descartes, *L'Homme* 2. English translations are my own.

77 Ibid. 2.

78 Ibid. 13.

79 Ibid. 57.

80 Ibid. 58.

81 Ibid. 58–59.

82 Ibid. 60.

83 Ibid. 4–5.

84 Ibid. 106–7.

85 Ibid. 131.

86 Thomas Hall (*L'Homme* xxix) cautions: "His view of the body is correctly termed 'mechanistic,' provided we understand what Cartesian mechanics entailed. Its major assumption is that the physical properties of bodies are reducible to the properties of three elements or variants of matter—specifically, to the shapes and motions with which the particles of the elements had originally been endowed. . . . The swirling

vortexes of the cosmos—and the water droplets of the rainbow—were, for Descartes, corpuscular machines."

87 Cited in Caverni 1:198.

88 For more on Borelli's biography see the preface by Carlo Giovanni, general superior of the Scuole Pie (Rome), included in *De motu animalium*; Barbieri's *Notizie istoriche*; Fisch's article on the Investiganti (536); and various references to him in Caverni, especially vol. 1. These are the sources I have used for the description of his life.

89 Note that the term *animal* includes humans. *Brute* often indicated what we now refer to as animals.

90 Caverni 1:189.

91 Borelli, *On the Movement of Animals* 222. We are fortunate to have this English translation by Paul Macquet of the 1743 edition.

92 Ibid. 1–2.

93 Ibid. 6.

94 Ibid. Carlo Giovanni, "Preface to the Reader" 4.

95 Ibid. 7.

96 Ibid. 8.

97 Ibid., 232–33.

98 Ibid. 368.

99 Ibid. 236.

100 Ibid. 237.

101 Ibid. 285.

102 Ibid. 318–20.

103 Ibid. 283–85.

104 Ibid. Giovanni, "Preface" 4.

105 Dini 134.

Five: Vico's Monstrous Body

1 Douglas 70–71. Douglas follows Marcel Mauss in maintaining that "the human body is always treated as an image of society and that there can be no natural way of considering the body that does not involve at the same time a social dimension." For example, "The relation of head to feet, of brain and sexual organs, of mouth and anus are commonly treated so that they express the relevant patterns of hierarchy. . . . [T]he same drive that seeks harmoniously to relate the experience of physical and social, must affect ideology. Consequently, when once the correspondence between bodily and social controls is traced, the basis will be laid for considering co-varying attitudes in political thought and in theology."

2 See Parel, especially chapters 5, 6, and 9, for a similar approach to Machiavelli's political theory.

3 Hobbes, preface: "Lastly the *pacts* and *covenants,* by which the parts of this body politic were at first made, set together, and united, resemble that *fiat,* or the *let us make man,* pronounced by God in the creation."

4 Ibid. preface.

5 Ibid. 237. "Amongst the infirmities therefore of a commonwealth, I will reckon in the first place, those that arise from an imperfect institution, and resemble the diseases of a natural body, which proceed from a defectuous procreation." Allowing citizens to read ancient works of political philosophy is likened to *hydrophobia,* the falling-down disease; and so forth.

6   Ibid. preface.

7   Ibid. 244.

8   See Dini 133.

9   Vico, *On the Most Ancient Wisdom* 48.

10 Ibid. 48–49.

11 Adams 46–47.

12 Fisch 552.

13 Ibid. 546.

14 Ibid. 548.

15 Fisch remarks that "it is not too much to say that he lived under the shadow of the Investigators." Fisch also provides a convenient comparative chronology: "Of the ablest resident members of the Academy in its most flourishing period, Bartoli died when Vico was eight, Cornelio when he was sixteen, Di Capoa when he was twenty-seven, Franceso d'Andrea when he was thirty, and Porzio when he was fifty-five. When the Academy finally ceased in 1737, Vico was nearly seventy" (552).

16 Ingegneri 45.

17 Dini 111.

18 Cited in Dini 113.

19 Fisch 539–40.

20 Di Capoa 112.

21 Ibid. 115.

22 Ibid. 41–42.

23 Ibid. 60.

24 Ibid. 88.

25 Fisch 522.

26 Di Capoa 135.

27 Sévérin 61.

28 Ibid. 3.

29 Ibid. 53: "Mais la molesse des habitans d'icelle est si importune qu'elle corrompt; comme faisoyent les Syrenes, les bons Medecins, qui n'ont pas la vigueur de resister, car ils n'osent proposer riend de desagreable ou de dangereux, ou pour se mettre hors de danger d'ignominie apprehendants quelques sinistres évenements, que puisse empescher les malades de commettre quelque faute: car combine y a-t'-il û de Medecins honorables qui si sont oüys nommer oyseaux de mauvaise augure & songeurs . . . ?"

30 Di Capoa 74.

31 Ibid. 135.

32 Ibid. 83.

33 Fisch 537.

34 Barbieri 161.

35 Ibid. 162.

36 English citations come from *The Autobiography of Giambattista Vico*, trans. Fisch and Bergin (1944), 133. When the Fisch and Bergin translation is inadequate or inappropriate I cite from the Italian version, providing my own English translation. The edition I used is *L'Autobiografia, il carteggio e le poesie varie*, ed. Croce, *Opere* (Bari: Laterza, 1911) vol. 5. A few references indicate the later, 1929, edition that Nicolini added to.

37 Barbieri 162.

38 See Torrini for more on Cornelio and his connection to Vico.

39 Royal Society of London 579.

40 Ibid. 577.

41 Barbieri 164.

42 Vico, *Autobiography*, trans. Fisch and Bergin, 153.

43 Dini 131. Porzio explicitly cites Descartes's *L'Homme* and shows in-depth familiarity with *Le Monde ou Traité de la lumière*, the *Discourse on Method and Essays*, the *Principia philosophiae*, and the three volumes of *Letters* published between 1657 and 1667 by Clerselier.

44 Vico, *L'Autobiografia, il carteggio e le poesie varie*, 37.

45 Vico, *Opere*, ed. Nicolini (1953) 969.

46 Barbieri 166.

47 Nicolini refers to the *De Antiquissima* as Vico's *Liber metaphysicus*. Vico's idea at one time was to derive a never written *Liber moralis* "precisely from the cosmogony developed in the lost *Liber physicus*." As Vico himself says: "For the reasons set forth in the little book which he afterwards published [*De aequilibrio*, the *Liber physicus*], Vico now undertook to ground his physics on a suitable metaphysics [*De antiquissima*]" (*Autobiography*, trans. Fisch and Bergin, 151).

48 Vico, *Autobiography*, trans. Fisch and Bergin, 138.

49 "con mente metafisica le fisiche cose." Cited in Dini 121.

50 Vico, *Autobiography*, trans. Fisch and Bergin, 121.

51 Doria, *Dialoghi* preface.

52 Ibid.

53 Ibid. 76. "La Geometria vale a darci la conoscenza del vero unico, ed in astratto, e serve a formare il Metafisico, ed in consequenza di ciò a far si, che la mente umana faccia idea della Giustizia, che è una, ed a far formare ancor di tutti più particolari, la giusta idea; alla perfine serve a formare il vero uomo."

54 Vico, *On the Most Ancient Wisdom* 44.

55 Doria 76. "ed hanno ridotto prima la Matematica alla Chimera, e poscia hanno fatto di quella una Fisica, & un Meccanica; e quindi è poi, che lo stesso disordinato metodo seguendo, nello studio della Filosofia, delle leggi, e delle morali, hanno gli uomini, in queste importantissime facoltà, tanta mostruosità di diverse sentenze seguite."

56 Vico, *On the Most Ancient Wisdom* 44.

57 Dini 122.

58 Vico, *On the Most Ancient Wisdom* 68.

59 Letter to Francesco Estevan, 1729, in *Opere*, ed. Nicolini (1953) 135.

60 Vico, *L'Autobiografia, il carteggio e le poesie varie* 21.

61 Vico, *Opere*, ed. Nicolini (1953), 131.

62 Vico, letter to Gherardo Degli Angioli, 1725, in *Opere*, ed. Nicolini (1953) 121.

63 Ibid. 237.

64 Ibid. 130.

65 Ibid. 21, note 6: "Il che, tuttavia, non gl'impedi' di condurre in gioventu' studi di medicina."

66 Vico, *L'Autobiografia, il carteggio e le poesie varie* 17.

67 Ibid. 963.

68 Vico, *On the Wisdom* 77.

69 Adams 41. At the time of writing the poem Vico was listening "to the mighty torrent of Dante." His appreciation for the heroic language of Dante supports my belief that this poem should be read as a work of natural philosophy in the model of the medieval philosopher-poets and Lucretius.

70 English translations of the poem are my own. The poem can be found in *Opere,* ed. Nicolini (1953) 159–63.

71 Vico, *Opere,* ed. Nicolini (1953) 963.

72 Vico, *L'Autobiografia, il carteggio e le poesie varie* 68.

73 Vico, *Opere,* ed. Nicolini (1953) 95.

74 Vico, *L'Autobiografia, il carteggio e le poesie varie* 38.

75 Ibid. 74–75.

76 I agree with Nicolini in his skeptical assessment of Villarosa's account. See Nicolini's "Gli ultimi anni," in *Opere* (1953) 100–106 for a more likely story.

77 Vico, *L'Autobiografia, il carteggio e le poesie varie* 3.

78 Vico, *Opere,* ed. Nicolini (1953) 241.

79 Vico, *The New Science,* trans. Fisch and Bergin, §919. References to the *New Science* indicate the paragraph number rather than the page.

80 Lilla 182.

81 Vico, *On Humanistic Education* 2:70.

82 Ibid. 6:127.

83 Vico, *On the Most Ancient Wisdom* 87.

84 Vico, *On Humanistic Education* 1:45.

85 Vico, *On the Most Ancient Wisdom* 65, 61.

86 Ibid. 65–66.

87 Ibid. 60.

88 Ibid. 70.

89 Vico, *On Humanistic Education* 6:130.

90 Ibid. 3:74, 75.

91 Ibid. 3:75.

92 Ibid. 6:133.

93 Ibid. 3:79.

94 Ibid. 3:81.

95 See the notes by translators Pinton and Shippee, *On Humanistic Education.*

96 Compare, for example, the similar story, p. 61 in Di Capoa's *Parere.*

97 Vico, *On the Most Ancient Wisdom* 55.

98 Ibid. 88, 89.

99 Vico, *L'Autobiografia, il carteggio e le poesie varie* 33–34.

100 Vico, *Opere,* ed. Nicolini (1953) 189–91. A *sorite* is a form of argument having several premises and one conclusion, capable of being resolved into a chain of syllogisms, the conclusion of each of which is a premise of the next (*Random House Dictionary*).

101 Ibid. 189.

102 Vico, *L'Autobiografia, il carteggio e le poesie varie* 133. "Fu la sua statura delle mediocri, l'abito del corpo adusto, il naso aquilino, e gli occhi vivi e penetranti, dal cui fuoco avrebbe ognuno potuto facilmente comprendere qual fosse la forza e l'energia di sua vigorosa mente. Contribuí alla sublimitá e speditezza dell'ingegno il suo collerico temperamento." The interested reader may also consult the one extant portrait of Vico, a copy of a lost original, reproduced in the *Autobiography,* trans. Fisch and Bergin.

103 Stelluti 102.

104 Ibid. 97, 30.

105 "Amava i suoi con eccesso di tenerezza, contento piuttosto di una respettosa amicizia che d'un servile timore." *L'Autobiografia, il carteggio e le poesie varie* 133. The last sentence is quoted in English in Adams, *Life and Writings* 182.

106 Vico, *L'Autobiografia, il carteggio e le poesie varie* 3.

107 *Adust* indicates "dried or darkened as by heat; burned, scorched; gloomy in appearance or mood" (*Random House Dictionary*).

108 Stelluti 21.

109 Ibid. 22.

110 Vico *L'Autobiografia, il carteggio e le poesie varie* 120.

111 In *Opere*, ed. Nicolini (1953) 104–5.

112 Vico, *Autobiography*, trans. Fisch and Bergin, 111.

113 Vico, *Opere*, ed. Nicolini (1953) 153.

114 Ibid. 91.

115 Ibid. 105.

116 Mandeville xii.

117 Ibid. xiv.

118 Ibid. 54.

119 Ibid. 106–7.

120 Vico, *Autobiography*, trans. Fisch and Bergin, 181.

121 Vico, *L'Autobiografia, il carteggio e le poesie varie* 76.

122 Correspondence with Francesco Saverio Estevan, dated January 24, 1729. See *L'Autobiografia, il carteggio e le poesie varie* 205–6. The fact that Vico included this oration among a sampling of his works in response to Estevan's request would indicate that in spite of the author's dismissal of the rhetorical piece as "una operuccola fatta per passatempo [bagatelle of a work, made to pass time]" (203), he must have been proud of its impeccable eloquence and considered it an important meditation on the heroic character of the second age.

123 Vico, *Autobiografia. Seguita da una scelta di lettere, orazioni e rime*, ed. Mario Fubini (Torino: Einaudi, 1977), 155. Page numbers refer to this edition. A paraphrase provided by Fubini in his introduction to the oration: "In Morte di Donn'Angela Cimmino Marchesa della Petrella. Orazione premessa alla miscellanea poetica dal Vico stesso promossa e curata per l'occasione (1727)." The oration can also be found in Nicolini's *Opere*, pp. 1009–31. I use the spelling adopted in the body of the oration: "Angiola Cimini."

124 The description that Vico gives this work in his catalog is as follows: "Il cui argomento essendo che questa valorosa donna nella sua vita insegnò il 'soave austero' della virtú, a proposito della materia l'auttore ha unito il delicato de' sensi greci e 'l robusto dell'espressioni all'aria grande latina e gli ha condotti coi colori della italiana favella." *L'Autobiografia, il carteggio e le poesie varie* 87.

125 Adams 180.

126 Vico, "In Morte," ed. Fubini, 172.

127 Ibid. 172.

128 See Dini 49 ff.

129 Vico, *Scritti storici*, in *Opere*, ed. Nicolini (1939) 6:389–40.

130 Ibid. 400.

131 Vico, "In Morte," ed. Fubini, 174.

132 Ibid. 175.

133 Ibid. 169.

134 Ibid. 170.

135 Vico, *The New Science* §191.

136 Ibid. §405.

137 Vico, "In Morte," ed. Fubini, 176.

138 Ibid. 188.

139 Vico, *Autobiography,* trans. Fisch and Bergin, 150.

140 Ibid. 151. The exact cause for fever is as follows: "May not fevers be caused by air in the veins moving from the heart at the center to the periphery and distending, more than is compatible with good health, the diameters of blood vessels clogged at the opposite, or outer, end? On the other hand, may not malignant fevers be motion of air in the blood vessels inward from without, likewise distending more than is compatible with good health the diameters of vessels clogged at the opposite, or inner, end?" (150–51).

141 Vico, "In Morte," ed. Fubini, 191.

142 Vico, *On the Most Ancient Wisdom* 79. For more on *conatus,* see Badaloni 19–25.

143 Vico, *On the Most Ancient Wisdom* 86.

144 Ibid. 80, 85.

145 Ibid. 83, 97.

146 Ibid. 110.

147 Ibid. 92.

148 "First Response," *On the Ancient Wisdom of the Italians* 133. See especially pp. 142–43 for the disputation on the nerves, blood, and free will.

149 Vico, *New Science,* §1407, §1411. From the "Practic" appended to the English edition.

150 Ibid. §1409.

151 Ibid. §1409, §195, §170.

152 Ibid. §369.

153 See Dini 84 for Porzio's discussion on this topic.

154 Vico, *New Science* §1410.

155 Ibid. §504.

156 Ibid. §340.

157 Vico, *On Humanistic Education* 3:74.

158 Vico, *New Science* §341.

159 Ibid. §376, §340, §375.

160 Ibid. §377.

161 Ibid. §504, §1098.

162 Ibid. §520.

163 Ibid. §692.

164 Ibid. §1405, §1411.

165 Ibid. §540.

166 Hercules' second encounter following the Nemean lion was with the Hydra, a hideous, many-headed creature, half-sister to the Nemean lion. His third and fourth labors involved the capture of wild animals, testaments to his superior hunting skills. The fifth landed him in dung, cleaning up the Augeian stables; the sixth was to shoo away the man-eating Stymphalian birds; the seventh, to capture and bring back the Cretan bull, father to the Minotaur. The eighth labor imposed on him by King Eurystheus was to fetch the mares of Diomedes, who were raised on human flesh. The ninth involved stealing a belt from Hippolyte, queen of the Amazons; on the way back he subdued a sea monster in return for a girl. The tenth labor was to steal the cattle of Geryon, the triple-headed monster-king of Erytheia. (He stopped on the way home to kill the giant Cacus.) The eleventh labor was to bring back to the king the apples of the Hesperides, a garden tended by nymphs, with the aid of a hundred-headed snake named Ladon. And his final labor was to bring up Cerberus, the hound with three (or

fifty) heads and a snake's tail, brother of the Hydra and the Nemean lion, who guarded the gates of Hades.

167   C. M. Bowers, cited in duBois, *Centaurs and Amazons* 57, 58.

168   Ripa 108–9, 121.

Six: Monstrous Metaphor

1   Le Bon, "Psychologie des foules," cited in Freud 11.

2   E. Tesauro 696.

3   Alciati, *Emblemata* (Padua, 1621), xv. See the introduction to volume 1 of the Daly edition for a brief history of Alciati's life and works as well as for bibliographical information. Praz, *Studies in Seventeenth-century Imagery*, gives an excellent introduction to the history and aesthetics behind emblems as well as tracing the development of several of Alciati's emblems. See chapter 1: "Emblem, Device, Epigram, Conceit," pp. 11–54.

4   Ibid. Translation from *Emblems in Translation*. The Latin original:

> *Sirenes*
> Absque alis volucres, & cruribus absque puellas,
> Rostro absque & pisces, qui tamen ore canant,
> Quis putat esse ullos? iungi haec Natura negavit:
> Sirenes fieri sed potuisse docent.
> Illicium est mulier, quae in piscem desinit atrum,
> Plurima quod secum monstra libido vehit.
> Aspectu, verbis, animi candore trahuntur,
> Parthenope, Ligia, Leucosiaue viri.
> Has Musae explumant, has atque illudi Ulysses:
> Scilicet est doctis cum meretrice nihil.

5   Céard and Margolin, *Rébus* (Paris: Maisonneuve et Larose, 1986).

6   E. Tesauro 694.

7   *Meraviglia* is a complex term whose many semantic connotations make it difficult to translate with one equivalent. Other possibilities include: *wonderment, amazement, astonishment, surprise, awe. Wonder,* I think, comes closest since it expresses curiosity (as in "to wonder at something"), positive marvel (as in "wonderful"), and a prodigious quality (as in "the wonders of nature").

8   Aristotle, *Metaphysics* 982b, 12ff.

9   Plato, *Theaetetus* 155d.

10   Aristotle, *Rhetoric* 1371a, 31.

11   Pallavicino, *Trattato dello stile* 197.

12   As is suggested by both its Latin root and the textual context in which it appears in the seventeenth century.

13   Aristotle, *Rhetoric* 1412b, 1411b.

14   Praz, *Studies in Seventeenth-century Imagery* 15.

15   Campanella would have us study instead the living book of the world, a theme he explores in his sonnet "Modo di Filosofare." *Tutte le opere* 1:18.

16   For the connection between monstrous images and their uses in imprinting memories, see Yates, *The Art of Memory;* Spence; and Dieckmann.

17   See Fagiolo dell'Arco and Carandini; and especially Krautheimer.

18   Peregrini 26. See Fantuzzi 6:331–33 for biographical details. Variants of his name

include Peregrini, as in most of his published works; Pellegrini also appears in one text (*Fonti dell' Ingegno ridotti ad Arte*), and an academician refers to him posthumously by that spelling. The entry in Fantuzzi appears under the name Pellegrino. I have chosen, like Ezio Raimondi, to use what seems to be the most frequently adopted spelling by Sig. Matteo himself: Peregrini.

19 Descartes, *Les Passion de l'âme* 198–200.

20 E. Tesauro 89.

21 Descartes, *Les Passions de l'âme* 158.

22 Ibid. 199.

23 Ibid. 200–201.

24 "Baconian" perhaps does not take into account Bacon's own distinctions and contradictions. Compare: "Our experiments we take care to be, as we have often said, either experimenta fructifera or lucifera: either of use or of discovery: for we hate impostures and despise curiosities" (*Sylva sylvarum*, century 6, opening passage before item 501); and another passage in which he attributes the failure of experimental science so far to the fact that men have "sought out experiments for the sake of gain and not of knowledge" (*Phaenomena universi* in *Works* (1863 ed.) 7:230.

25 In *Valerius Terminus I*, 1603 fragment published posthumously in 1734.

26 Stigliani, *Epistolario* (1636), quoted in Franco Croce 106: "Un tempo i lettori si contentavano d'una lettura non cattiva, poi volsero eccellenza, appresso desiderarono meraviglie, ed oggi cercano stupori, ma dopo avergli trovati, gli hanno anco in fastidio ed aspirano a trasecolamenti e a strabiliazioni."

27 Ibid. 106. "The great lack of appetite that pregnant women usually have sometimes degenerates into a faulty desire to eat charcoal and lime mortar or chalk or like filth, in the same way the extreme satiety of our readers, from their desire to vary their pastures has converted into a crazy appetite to read foolishnesses."

28 Pallavicino 197; Peregrini 166.

29 The term *pellegrino* is a figurative usage of the word that signifies "a pilgrim," hence by extension something that is "wandering, roaming, or wayfaring" and equally something that is "foreign, or alien." The term thus ends up in literary usage as signifying "strange, outlandish, or uncommon," and this is the most common usage of *pellegrino*, in fact, that we find in *seicento* literary texts.

30 Franco Croce's book *Tre momenti del barocco letterario* is an excellent presentation and analysis of this debate.

31 D. Bartoli, "Uomo di lettere difeso ed emendato" 127, 132.

32 Peregrini 219.

33 Ibid. 201, 215, 161, 222, 256, 149, 153.

34 Ibid. 231–32. Bartoli also devotes an entire section of his "Uomo di lettere" (114–21) to "the madness of many who, wanting to appear learned, have themselves published with ignorant printers."

35 Peregrini 199, 202, 202.

36 Ibid. 196.

37 Ibid. 42.

38 Ibid., dedication.

39 Ibid. 206, 249, 162, 199.

40 I have chosen in this study to take these equivalences at face value and not to examine the differences between the various technical terms that make up the *seicento* rhetorical lexicon. Peregrini himself in the title of his treatise indicates that *acutezze* are indiscriminately referred to as "spiriti," "vivezze," and "concetti": "Delle Acutezze che

altrimenti Spiriti, Vivezze, e Concetti, Volgarmente si appellano." Some would argue that *arguzia* refers to wit as a rhetorical strategy for defense and self-defense, a situational response in a rhetoric of manners, while *argutezza* should more properly be understood as an intellectual quality of brilliance or cleverness. Vico, for example, narrates (in his *Autobiografia*, ed. Fubini [1977], 10) that he indulged "in the most corrupt manners of modern poetizing" purely as "an exercise of the wit in works of witticisms [un esercizio d'ingegno in opere d'argutezza] which delights itself solely with the false, in extravagant appearance, and in surprising the listeners' direct expectations" (10). In fact, Vico supports the systematic practice of witticisms in order to develop "young people's minds [ingegni], which are overly refined and rigidified by the study of metaphysics" (10). Thus Peregrini's use of *acutezza* would be (literally) a pointedness, a sharp stab at one's opponent, whereas Tesauro's emphasis on *arguzia* brings out the human capacity for sublime wit as afforded by the superior "ingegno." These sorts of distinctions are obviously of tremendous consequence and would require extensive treatment.

41   See Raimondi's article "Una data da interpretare" in *Letteratura barocca* 51–75.

42   Peregrini 166, 45, 57, 61, 56, 74, 75.

43   Ibid. 202.

44   How should *ingegno* be translated into English? One could say the "wit" of the speaker, in contrast to the "witticism" itself. "Ingeniousness" is awkward, but perhaps the most accurate rendering.

45   Peregrini 165, 42, 206.

46   Ibid. 164, 170, 180, 182, 183.

47   Ibid. 248, 166.

48   Fantuzzi 6:333.

49   L. Tesauro 9.

50   E. Tesauro 1, 536, 584: "On the contrary, the human mind participates in the Divine; which with the same Divinity dwells in the swamps, & in the Stars: & from the most sordid nettle-tree makes the most Divine of the Corporeal Creatures."

51   Gracián, *Agudeza y arte de ingenio* 1:51. "Si el percibir la agudeza acreditada di águila, el producirla empeñará en ángel; empleo de querubines, y elevación de hombres, que nos remonta a extravagante jerarquía." There is no modern English translation, but a French edition exists.

52   E. Tesauro 96, 83.

53   Ibid. 16–17, 122. This is Tesauro's version of the Aristotelian dictum that "to change is also pleasant: change means an approach to nature, whereas invariable repetition of anything causes the excessive prolongation of a settled condition: therefore, says the poet, Change is in all things sweet" (*Rhetoric* 1371a, 26–30).

54   Gracián, *Agudeza y arte de ingenio*: "Consiste, pues, este artificio conceptuoso, en una primorosa concordancia, en una armónica correlación entre dos o tres cognoscibles extremos, expresada por un acto del entendimiento" (1:55); "Siempre el hallar correspondencia entre los correlatos es fundamento de toda sutileza" (1:97).

55   E. Tesauro 446, 452.

56   Aristotle, *Rhetoric* 1412a, 1412b, 1414a.

57   Ibid. 1404b, 1408b, 1410b.

58   E. Tesauro 82, 293, 449.

59   Ibid. 364. As absurd as this cybernetic creature composed out of cooking utensils may sound, at least two graphic versions of it exist. See *Effetto Arcimboldo* 113–14. While these composite creations are generally presented as allegories of their subject, inter-

estingly enough, in the same series (*Humana victus instrumenta*), there is another figure composed out of agricultural implements whose title reads "The true portrait of a stupendous monster seen recently by peasants in a well-cultivated field, 1567," *Effetto Arcimboldo*, 115. Perhaps the overworked peasants were haunted by a monstrous phantasm of their labor?

60 There is some question as to the actual date of composition of *Il Cannocchiale*. Ezio Raimondi argues convincingly that it must have been written in a youthful period, probably in the 1630s, when Tesauro was most influenced by Marino. (See "Una data da interpretare," in *Letteratura barocca* 51–75.) Franco Croce (*Tre momenti* 154) argues that the work is a product of mature thought, and although begun in the 1630s, it must have been completed in the 1650s. I would agree with Croce on this point.

61 Nicolini, *Sulla vita civile* 14–15.

62 Ibid., cited on 14.

63 Ibid., cited on 15.

64 "Il più grande oratore sacro del Seicento italiano," Raimondi, *Trattatisti* 653.

65 Paolo Segneri, *Il Cristiano instruito nella sua legge* (1686), "Ragionamento XXXI, In detestazione delle commedie scorrette," reprinted in Raimondi, *Trattisti* 750–62, 750.

66 Ibid. 750, 751, 654.

67 Ibid. 759: "La virtù de' cristiani è una virtù di croce, di contradizione e di violenza . . . il regno de' cieli non si darà se non a chi l'ottiene a forza e l'espugna con l'arme in mano. 'I violenti se ne impadroniscono.' "

68 Ibid. 755, 757, 752, 757.

69 Ibid. 760: "Non è un incantesimo? non è un fascino? non è una fattucchieria? E poi chi sono costoro, da cui vi lasciate così stravolgere? Quali sono, dirò così, quali sono que' maghi che vi hanno ammaliata la fantasia? . . . Sono ciurma di gente vile."

70 Ibid. 762.

71 Croce, "I Predicatori italiani del seicento e il gusto spagnuolo," *Saggi* (1948) 155–81.

72 Ibid. 163–64.

73 Ibid. 168.

74 E. Tesauro 502–3. Croce (*Saggi*) concurs that "the witty turns were very well received not only by the educated public of the academies and the elegant members of courts, but also by the rabble: the popularity of those witticism- and metaphor-mongering preachers was widely based" (173). Even more to Croce's perplexity (whose aesthetic taste ran quite contrary to the people of Baroque Italy), "those figures of speech often made torrents of tears gush forth" (163).

75 E. Tesauro 501.

76 Ibid. 539.

77 Ibid. 507, 539.

78 Ibid. 501.

79 All the quotations in this section come from E. Tesauro 511–17.

80 Ibid. 512. "Nella tiepida spiaggia del Mar Tirreno sollazzavano tre bellissime Ninfe, che lusinghevolmente invitando i Naviganti à volger le vele al loro lido, promettendolo un nido delle Gratie & degli Amori; patria dell'allegrezza, porto della quiete, paradiso delle delitie & de' piaceri: con triplicate legami de' vaghi aspetti, delle corde canore, & delle voci soavi, cantando incantavano, & incantando incatenavano inguisa quegli' nfelici; che scordati de' loro affari, & dell'amata patria ove tendeano; non mirando altra Stella che que' begli occhi, ne seguendo altro vento, che l'aria piacevole di quelle armoniose voci; à remi, à vele, à volo con lieta celeusma, spignean le navi carche di ricche merci & di novelle speranze à quella riva. Ma ecco, che apena approdati, altro non veggendosi'

ntorno che un'horrido deserto, e nudi scogli, di nude ossa, e d'insepulti cadaveri dis-
seminati, quì un teschio pendente, là un busto corroso, altrove un tronco ancor semi-
vivo: dal terror della Morte pria che dalla Morte abbattuti, lasciavano in preda à quelle
Fiere canore, le navi, le merci, e alfin la vita."

81   Translated as dragons, jackals, or monsters in English-language Bibles.
82   Ibid. 517: "Finalmente si de' conchiudere che, sicome il prudentissimo Ulisse, solcando
il Mare per giugnere velocemente alla sospirata sua Patria; si sè strettamente legare
all'Arbore della Nave: & incerò le orecchie a' Remiganti per non lasciarsi frastornare
dal suo diritto camino dalla voce soave delle crudeli Sirene: così l'Anima saggia tena-
mente si de' attenere all'Arbore di Santa Croce: & chiudendo le orecchie alle canore
lusinghe della Voluttà, aprirle alla divina Voce, che la chiama al Porto del Cielo, dove
sono i veri piaceri, e dove son dirizzati i nostri pensieri."
83   Ibid. 96. "che più non pare Huom terreno; ma un celestial Nume nell'Arte sua."
84   José Francisco de Isla, *Historia del famoso predicator Fray Gerundio de Campazas, alias
Zotes* (1758); quoted in Croce, *Saggi* 180.
85   E. Tesauro 503: "ugualmente i piccoli, e i grandi; i nobili, & i plebei."
86   Ibid. 502.
87   Croce, *Saggi* 178–81.
88   E. Tesauro 1.
89   Both a poison and a cure, the siren is thus an example of a pharmakon. This characterization links sirens and monsters with the scapegoat, a relation I have hinted at
throughout this book.
90   Horkheimer and Adorno xiii.
91   Ibid. 32.
92   Ibid. 13.
93   Ibid. 33.
94   E. Tesauro 649.
95   Horkheimer and Adorno 31.
96   Luigi Vigliani goes so far as to hail Tesauro's historiography as an exceptional precursor
to full-blown Romantic historiography (as the realization of the spirit of the nation) for
the fact that unlike his contemporaries who generally were nomadic, cynical, and took
pride in forgoing political loyalties, Tesauro was anchored by his Piedmontese roots and
undying devotion to his prince.

BIBLIOGRAPHY

Adams, H. P. *The Life and Writings of Giambattista Vico*. London: George Allen and Unwin, 1935.

Agrippa Von Nettesheim, Henry Cornelius. *Occult Philosophy or Magic [Translation of Book 1 of De occulta philosophia, 1533]*. Trans. W. F. Whitehead, using 1651 English translation. Reprint of Hahn and Whitehead edition (Chicago, 1898). New York: AMS Press, 1982.

Alberi, Eugenio, ed. *Tesoro di Prose Italiane*. Florence: Società Editrice Fiorentina, 1841.

Alciati, Andrea. *Emblemata*. Padua, 1621.

——. *Emblems in Translation*. Ed. Peter Daly. 2 vols. Toronto: U of Toronto P, 1985.

Aldrovandi, Ulisse. *Monstrorum historia*. Bologna, 1642.

Aristotle. *The Basic Works of Aristotle*. Ed. Richard McKeon. New York: Random House, 1941.

Ashworth, William B. "Natural History and the Emblematic World-View." In *Reappraisals of the Scientific Revolution*, ed. David L. Lindberg and Robert S. Westman. Cambridge: Cambridge UP, 1990.

Augustine. *City of God*. Trans. Henry Bettenson. Middlesex: Penguin Books, 1972.

Bacon, Francis. *New Atlantis*. New York: Odyssey Press, 1937.

——. *Selections*. Ed. M. T. McClure. New York: Scribner's, 1928.

Badaloni, Nicola. *Introduzione a Vico*. Rome-Bari: Laterza, 1984.

Baltrusaitis, Jurgis. *Le miroir: révélations, science-fiction et fallacies*. Paris: Seuil, 1979.

——. *Réveils et Prodiges: Le gothique fantastique*. Paris: Armand Colin, 1960.

Bann, Stephen. "Le Sérieux de l'emblème." In *Figues du Baroque*, ed. Jean-Marie Benoist. Paris: Presses Universitaires de France, 1983.

Barbieri, Matteo. *Notizie istoriche dei mattematici e filosofi del regno di Napoli*. Naples: V. Mazzola-Vocola, 1778.

Barkan, Leonard. *The Gods Made Flesh: Metamorphosis and the Pursuit of Paganism*. New Haven: Yale UP, 1986.

Barnes, Jonathan. *Early Greek Philosophy*. Middlesex: Penguin Books, 1987.

Bartoli, Daniello. "Della ricreazione del Savio." In *Antologia della prosa scientifica italiana del seicento*, ed. Enrico Falqui. Florence: Vallecchi, 1943.

——. "Uomo di lettere difeso ed emendato." In *Prose scelte di Daniello Bartoli e Paolo Segneri*, ed. Mario Scotti. Turin: U.T.E.T., 1967.

Bartoli, Sebastiano. *Thermologia Aragonia* [sive Historia naturalis Thermarum In Occidentali Campaniae ora inter Pausilippum, & Misenum Scatentium, iam Aeui iniuria deperditarum, & Petri Antonii Ab Aragonia Studio, ac munificentia restitutarum, ubi eruditè differitur de Pyrosophiae, & Hydrosophiae Arcanis, origine Fluminum, incremento Nili, Aestu Maris, exalationibus terrae, ac insuper de Calore, & Lce non vulgariter philosophatur, authore Sebastiano Bartolo In Neapolitano Gymnasio Philosophiae, & Anatomiae professore primario, Opus posthumum recensitum a Michaele Blancardo Philosoph. ac Medic. Doct. Authoris discipline Alumno]. Naples: Ex Typographia Novelli de Bonis, 1679.

Battisti, Eugenio. *L'antirinascimento*. Milan: Feltrinelli, 1962.

Bedani, Gino L. C. *Vico Revisited: Orthodoxy, Naturalism and Science in the Scienza Nuova*. Oxford: Berg Publishers, 1989.

Benoist, Jean-Marie, ed. *Figures du baroque*. Paris: Presses Universitaires de France, 1983.

Benzoni, Gino. *Gli Affanni della cultura intellettuale*. Milan: Feltrinelli, 1978.

Berti, Luciano. *Il Principe dello Studiolo. Francesco I dei Medici e la fine del Rinascimento florentino*. Florence: Editrice Edam, 1967.

Bethell, S. L. "Gracián, Tesauro, and the Nature of Metaphysical Wit." *Northern Miscellany of Literary Criticism* 1 (1953): 19–40.

Black, David W. *Vico and Moral Perception*. New York: Peter Lang, 1996.

Blanchet, Léon. *Campanella*. New York: Burt Franklin, 1920.

Blumenberg, Hans. *The Legitimacy of the Modern Age*. Ed. Thomas McCarthy. Trans. Robert M. Wallace. Cambridge, Mass.: MIT P, 1983.

Boaistuau, Pierre. *Histoires prodigieuses (1560)*. Paris: Club français du livre, 1961.

Bonet, Théophile. *Corps de Medecine et de Chirurgie*. Geneva: Chez Iean Anthoine Chovet, 1679.

Bonanni, Philippo. *Musaeum Kircherianum*. Rome, 1709.

Boorstin, Daniel J. *The Discoverers*. New York: Random House, 1983.

Borelli, Giovanni Alfonso. *On the Movement of Animals*. Trans. Paul Macquet. Berlin: Pringer-Verlag, 1989.

Brammall, Kathryn M. "Monstrous Metamorphosis: Nature, Morality, and the Rhetoric of Monstrosity in Tudor England." *Sixteenth Century Journal* 27.1 (1996): 3–22.

Branca, Giovanni. *Le machine*. Rome, 1629.

Brock, Bazon. "Immaculée Conception et machines célibataires." In *Junggesellenmaschinen/ Les machines célibataires*, ed. Jean Clair and Harald Szeeman. Venice: Alfieri, 1975.

Bronowski, J. *Magic, Science, and Civilization*. New York: Columbia UP, 1978.

Browne, Thomas. "Musaeum Clausum, or Bibliotheca Abscondita." *Selected Writings*. Chicago: U of Chicago P, 1968.

———. "Pseudodoxica Epidemica." *Selected Writings*. Chicago: U of Chicago P, 1968.

———. *Religio Medici (1642)*. 1881 ed. London: Macmillan, 1946.

Brumbaugh, Robert S. *Ancient Greek Gadgets and Machines*. New York: Thomas Y. Crowell, 1966.

Buci-Glucksmann, Christine. *La Folie de voir. De l'esthétique baroque*. Paris: Editions Galilée, 1986.

Campanella, Tommaso. *De Homine*. Trans. Romano Amerio. Vol. 13/14. Rome: Centro Internazionale di Studi Umanistici, 1960–61.

———. *Del senso delle cose e della magia*. Ed. Antonio Bruers. Bari: Laterza, 1925.

———. *La Città del Sole: Dialogo Poetico*. Trans. Daniel J. Donno. Berkeley: U of California P, 1981.

———. *Le Creature sovrannaturali*. Trans. Romano Amerio. Vol. 27. Rome: Centro Internazionale di studi umanistici, 1970.

——. "Narrazione della istoria sopra cui fu appoggiata la favola della ribellione." In *Il Supplizio di Tommaso Campanella*, ed. Luigi Firpo. Rome: Salerno Editrice, 1620.

——. *Tutte le opere*. Ed. Luigi Firpo. 2 vols. Turin: Mondadori, 1954.

Caprotti, Erminio. *Mostri, draghi e serpenti nelle silografie dell'opera di Ulisse Aldrovandi e dei suoi contemporanei*. Milan: Gabriele Mazzotta, 1980.

Cassirer, Ernst, Paul Oskar Kristeller, and John Herman Randall, eds. *The Renaissance Philosophy of Man*. Chicago: Chicago UP, 1948.

Cavell, Stanley. *The Claim of Reason: Wittgenstein, Skepticism, Morality, and Tragedy*. Oxford: Oxford UP, 1979.

Caverni, Raffaello. *Storia del metodo sperimentale in Italia*. 5 vols. Florence: G. Cavelli, 1891–90.

Cawson, Frank. *The Monsters in the Mind: The Face of Evil in Myth, Literature, and Contemporary Life*. Sussex: Book Guild, 1995.

Céard, Jean. *La Nature et les prodiges*. Geneva: Librairie Droz, 1977.

——. "Tératologie et tératomancie au XVIe siècle." In *Monstres et prodiges au temps de la renaissance*, ed. M. T. Jones-Davies. Paris: Jean Touzot, 1980.

——, ed. *La Curiosité à la renaissance*. Paris: Société d'édition d'enseignement supérieur, 1986.

Céard, Jean, and Jean-Claude Margolin. *Rébus de la Renaissance*. Paris: Maisonneuve et Larose, 1986.

Chartier, Roger. *Cultural History*. Oxford/Cambridge: Polity Press, 1988.

Chauvois, L. "William Harvey et l'Italie." (Conference proceedings) *Archivio italiano di scienze mediche, tropicali e di parassitologia* (1955): 383–400.

Cicero. *De Senectute, De Amicitia, De Divinatione*. Trans. W. A. Falconer. Loeb Classical Library ed. Cambridge, Mass.: Harvard UP, 1923.

Cipolla, Carlo M. *Faith, Reason, and the Plague in Seventeenth-Century Tuscany*. Trans. Muriel Kittel. New York: W. W. Norton, 1979.

——. *Fighting the Plague in Seventeenth-Century Italy*. Madison, Wis.: U of Wisconsin P, 1981.

Coffin, D. R. *The Italian Garden*. Cambridge, Mass.: Harvard UP, 1972.

Cohen, Jeffrey Jerome, ed. *Monster Theory: Reading Culture*. Minneapolis: U of Minnesota P, 1996.

Cook, Harold J. "Bernard Mandeville and the Therapy of "The Clever Politician." *Journal of the History of Ideas* 60.1 (1999): 101–24.

Cordier, Henri. *Les Monstres dans les légendes et dans la nature*. Paris, 1890.

Costa, E. *Ulisse Aldrovandi e lo studio bolognese nella seconda metà del secolo XVI*. Bologna, 1907.

Couliano, Ioan P. *Eros and Magic in the Renaissance*. Chicago: U of Chicago P, 1987.

Croce, Benedetto. *La Letteratura italiana: il seicento e il settecento*. Sansone ed. Bari: Laterza, 1957.

——. *Nuovi saggi sulla letteratura italiana del seicento*. Bari: Laterza, 1949.

——. *Saggi sulla letteratura italiana del seicento*. Bari: Laterza, 1948.

——. *Storia dell'età barocca in Italia*. Bari: Laterza, 1957.

——. *Tre momenti del barocco letterario italiano*. Florence: Sansoni, 1966.

Daston, Lorraine. "Marvelous Facts and Miraculous Evidence in Early Modern Europe." *Critical Inquiry* 18.1 (1991): 93–124.

Davidson, Arnold. "The Horror of Monsters." In *The Boundaries of Humanity: Humans, Animals, Machines*, ed. James J. Sheehan and Morton Sosna. Berkeley: U of California P, 1991.

Delcorno, Carlo. "Professionisti della parola: predicatori, giullari, concionatori." In *Tra storia e simbolo*. Vol. 46 of *Biblioteca di "Lettere Italiane" Studi e Testi*. Florence: Olschki, 1994.

Descartes, René. *L'Homme de René Descartes (1664)/Treatise of Man*. Thomas Steele Hall ed. (Facsimile edition originally published by Claude Clerselier, Paris: 1664). Cambridge, Mass.: Harvard UP, 1972.

——. *Les Passions de l'âme*. Ed. J.-M. Monnoyer. Paris: Gallimard, 1988.

——. *Oeuvres de Descartes*. Ed. Charles Ernest Adam and Paul Tannery. 12 vols. 2d. ed. Vol. 1. Paris: J. Vrin, 1956.

——. *Oeuvres et lettres*. Ed. André Bridoux. Paris: Gallimard, 1949.

——. *Oeuvres philosophiques*. Vol. 2. Paris: Garnier, 1967.

Di Capoa, Lionardo. *Del Parere* [del Signor Lionardo Di Capoa Divisato in otto Ragionamenti, Ne'quali partitamente narrandosi l'origine, e'l progresso della medicina, chiaramente l'incertezza della medesima si fa manifesta. Ultima edizione accresciuta di un'Indice copiosissimo, e delle postille nel margine. Dedicata All'Illustriss. ed Eccellentiss. Signore Il Signor D. Nicola Gaetano ell'Aquila D'Aragona Sesto Duca di Laurenzano, diciassettesimo Signore di Piedemonte, della Città, e Contea di Alise, della Baronie di Capriata, e di Alvignano, Capitano d'una Compagnia d'huomini d'armi del Regno di Nap., Principe di tutta la Famiglia Gaetano, ec]. 2 vols. Cologna (Naples): [n.p.], 1714.

Dick, Philip K. *Robots, Androids, and Mechanical Oddities*. Carbondale and Edwardsville: Southern Illinois UP, 1984.

Dieckmann, Liselotte. *Hieroglyphics: The History of a Literary Symbol*. St. Louis: Washington UP, 1970.

Dini, Alessandro. *Filosofia della natura, medicina, religione. Lucantonio Porzio (1639–1724)*. Milan: Franco Angeli Libri, 1985.

Dioguardi, Gianfranco. *Viaggio nella mente barocca*. Palermo: Sallerio, 1986.

Donato, Eugenio. "Tesauro's Cannocchiale Aristotelico." *Stanford Italian Review* 5 (Fall 1985): 101–13.

—— "Tesauro's Poetics: Through the Looking-glass." *MLN* 78 (1963): 15–30.

Donzelli, Giuseppe, Tomaso Donzelli, and Gio. Giacomo Roggieri. *Teatro Farmaceutico dogmatico, e spagirico* [del Dottore Giuseppe Donzelli napoletano, Barone di Digliola, Nel quale s'insegna una moltiplicità d'Arcani Chimici più sperimentati dall'Autore, in ordine alla sanità, con evento non fallace, e con una canonica norma di preparare ogni compositione, più costumata dalla Medicina Dogmatica: & una distinta, curiosa, e profittevole Historia di ciascheduno ingrediente di esse. Con l'aggiunta in molti luoghi del dottor Tomaso Donzelli figlio dell'autore, et in questa Quinta Impressione corretto, & accresciuto con un Catalogo dell'Herbe native del Suolo Romano del signor Gio: Giacomo Roggieri Roman]. Venice: Presso Paolo Baglioni, 1686.

——. *Teatro Farmaceutico dogmatico, e spagirico*. Venice: Appresso Andrea Poletti, 1713.

Doria, Paolo Mattia. *Dialoghi* [nei quali, rispondendosi ad un'Articolo de' Signori Autori degli Atti di Lipsia, s'insegna l'arte di esaminare una Dimostrazione Geometrica, e di dedurre dalla Geometria Sintetica la conoscenza del Vero, e del Falso; ed in conseguenza di ciò si esamina l'Algebra, ed i nuovi Metodi de' Moderni]. Amsterdam, 1718.

Douglas, Mary. *Natural Symbols: Explorations in Cosmology*. New York: Pantheon Books, 1982.

Drachmann, A. G. *Ktesibios, Philon, and Heron: A Study in Ancient Pneumatics*. Ed. Bibliotheca universitatis hauniensis. Vol. 4. Copenhagen: Ejnar Munksgaard, 1948.

duBois, Page. *Centaurs and Amazons.* Ann Arbor: U of Michigan, 1987.

Durbin and Rapp, eds. *Philosophy and Technology.* Dordrecht: D. Reidel, 1983.

Eamon, William. *Effetto Arcimboldo.* Milan: Bompiani, 1987.

———. "Plagues, Healers, and Patients in Early Modern Europe. *Renaissance Quarterly* 52.2 (1999): 474–87.

Einaudi, ed. *Storia d'Italia. Dalla caduta dell'Impero romano al secolo XVIII.* Vol. 2. Turin: Einaudi, 1974.

Eisenstein, Elizabeth L. *The Printing Revolution in Early Modern Europe.* Cambridge: Cambridge UP, 1983.

Evans-Pritchard, E. E. "The Notion of Witchcraft Explains Unfortunate Events." In *Reader in Comparative Religion,* ed. William A. Lessa and Evon Z. Vogt. New York: Harper and Row, 1937.

Fagiolo, Marcello, ed. *La città effimera e l'universo artificiale del giardino. La Firenze dei Medici e l'Italia del '500.* Rome: Officina Edizioni, 1980.

Fagiolo dell'Arco, Maurizio, and Silvia Carandini. *L'Effimero Barocco. Strutture della festa nella Roma del '600.* 2 vols. Rome: Bulzoni, 1977–78.

Falqui, Enrico, ed. *Antologia della prosa scientifica italiana del seicento.* Florence: Vallecchi, 1943.

Fantuzzi, Giovanni. *Notizie degli scrittori bolognesi.* 6 vols. Bologna: Stamperia di S. Tommaso d'Aquino, 1781–94.

Fassò, Guido. *Vico e Grozio.* Naples: Guida Editori, 1971.

Fedele, Francesco G., and Alberto Baldi, eds. *Alle origini dell'anthropologia Italiana: Giustiniano Nicolucci e il suo tempo.* Naples: Guida editori, 1988.

Ficino, Marsilio. *The Book of Life [De Vita Triplici].* Trans. Charles Boer. Irving, Tex.: Spring Publications, 1980.

Findlen, Paula. "Jokes of Nature and Jokes of Knowledge: The Playfulness of Scientific Discourse in Early Modern Europe." *Renaissance Quarterly* 43.2 (1990): 292–331.

———. "The Museum: Its Classical Etymology and Renaissance Genealogy." *Journal of the History of Collections* 1.1 (1989): 59–78.

———. *Possessing Nature: Museums, Collecting, and Scientific Culture in Early Modern Italy.* Berkeley and Los Angeles: U of California P, 1994.

Firpo, Luigi. *I processi di Tommaso Campanella.* Ed. Eugenio Canone. Rome: Salerno, 1998.

———, ed. *Il supplizio di Tommaso Campanella. Narrazioni—Documenti—Verbali delle torture.* Rome: Salerno Editrice, 1985.

Fisch, Max. "The Academy of the Investigators." In *Science, Medicine, and History: Essays on the Evolution of the Scientific Thought and Medical Practice, Written in Honour of Charles Singer,* ed. E. Ashworth Underwood. 2 vols. London: Oxford University Press, 1953.

Fischer, Jean-Louis. *De la genèse fabuleuse à la morphogénèse des monstres.* Paris: Société Française d'Histoire des Sciences et des Techniques, 1986.

Foresti, R. P. Antonio. *Mappamondo istorico.* Venice, 1691–94.

Foucault, Michel. *The Archaeology of Knowledge.* Trans. A. M. Sheridan Smith. New York: Pantheon Books, 1972.

———. *Madness and Civilization: A History of Insanity in the Age of Reason.* Trans. Richard Howard. Vintage Books ed. New York: Random House, 1988.

Freud, Sigmund. *Group Psychology and the Analysis of the Ego.* Trans. James Strachey. New York: Bantam, 1960.

Friedman, John Block. *The Monstrous Races in Medieval Art and Thought.* Cambridge, Mass.: Harvard UP, 1981.

Gareffi, Andrea. *Le voci dipinte. Figura e parola nel Manierismo italiano.* Rome: Bulzoni, 1981.

Garin, Eugenio. *La Cultura filosofica del rinascimento italiano.* Florence: G. C. Sansoni, 1961.

——. "Studi Italiani recenti sul Rinascimento." In *Magia, astrologia e religione nel Rinascimento,* ed. Accademi Polacca delle Scienze. Vol. 65. Wroclaw, Poland: Zaklad Narodowy Imienia Ossolinskich Wydawnictwo Polskiej Akademii Nauk, 1974.

——. "Vico and the Heritage of Renaissance Thought." In *Vico: Past and Present,* ed. Giorgio Tagliocozzo. Atlantic Highlands, N.J.: Humanities Press, 1981.

Garzoni, Tommaso. *La piazza universali dei professioni.* Venice: Michiel Maloco, 1665.

Gentilcore, David. "Contesting Illness in Early Modern Naples: Miracolati, Physicians, and the Congregation of Rites." *Past and Present* 148 (1995): 117–48.

Ghirardelli, Cornelio. *Cefalogia fisonomica* [divisa in dieci deche, dove conforme a' documenti d'Aristotile, e d'altri Filosofi naturali, con brevi discorsi, e diligenti osservationi si esaminano le Fisonomie di cento teste humane che intagliate si vedono in quest'Opera, dalle quali perpiù segni, e congetture si dimostrano varie inclinationi di Huomini, e donne di Cornelio Ghirardelli Bolognese, il Sollevato Academico Vespertino. Aggiontavi altretanti sonetti di diversi eccellenti Poeti, et Academici, ne' quali le prefate Fisonimie leggiadriamente racolgonsi. Et Additioni a ciascun Discorso dell'Inquieto Academico Vespertino]. Bologna: Presso gli Heredi di Evangelista Dozza e Compagni, 1670.

Giedion, Siegfied. *Mechanization Takes Command.* New York: Oxford UP, 1948.

Girard, René. *Violence and the Sacred.* Trans. Patrick Gregory. Baltimore: Johns Hopkins UP, 1977.

Godwin, Joscelyn. *Athanasius Kircher: A Renaissance Man and the Quest for Lost Knowledge.* London: Thames and Hudson, 1979.

Gorham, Geoffrey. "Mind-Body Dualism and the Harvey-Descartes Controversy." *Journal of the History of Ideas* 55.2 (1994): 211–34.

Gracián, Baltasar. *Agudeza y arte de ingenio.* Ed. Evaristo Correa Calderón. 2 vols. Madrid: Clásicos Castalia, 1969.

——. *Art et figures de l'esprit (Agudeza y arte del ingenio, 1647).* Trans. Benito Pelegrín. Paris: Seuil, 1983.

Grotius, Hugo. *Prolegomena to the Law of War and Peace.* Trans. Francis W. Kelsey. New York: Liberal Arts Press, 1957.

Guazzo, Francesco Maria. *Compendium Maleficarum.* Trans. E. A. Ashwin. Montague Summers, London, 1929 ed. New York: Dover, 1988.

Gusdorf, Georges. *La Révolution Galiléenne.* 2 vols. Paris: Payot, 1969.

Habermas, Jurgen. *Toward a Rational Society.* London: Heinemann, 1971.

Hacking, Ian. *Representing and Intervening.* Cambridge: Cambridge UP, 1983.

——. "Two Souls in One Body." *Critical Inquiry* 17.4 (1991): 838.

Harrison, Peter. "The Virtues of Animals in Seventeenth-Century Thought." *Journal of the History of Ideas* 59.3 (1998): 463–84.

Harroway, Donna. *Primate Visions.* New York: Routledge, 1989.

Harvey, William. *The Anatomical Exercises of Dr. William Harvey . . . Concerning the Motion of the Heart and Blood (1653).* Special limited facsimile edition originally published London: printed by F. Leach for R. Lowndes ed. [United States]: DevCom, 1987.

Hatzfeld, Helmut. "A Clarification of the Baroque Problem." *Comparative Literature* 1 (Spring 1949): 133–39.

Haydn, Hiram. *The Counter-Renaissance.* New York: Grove Press, 1960.

Hazard, Paul. *La Crise de la conscience européenne (1680–1715).* Paris: Boivin, 1934.

Hédélin, François. *Des Satyres brutes, monstres et démons.* Paris: Nicolas Buon, 1627.

Hickman, Larry, ed. *Philosophy, Technology, and Human Affairs*. College Station, Tex.: IBIS Press, 1985.

Hobbes, Thomas. *Leviathan*. Ed. Michael Oakeshott. Intro. by Richard S. Peters. New York: Macmillan, 1962.

——. *Leviathan*. New York: Liberal Arts Press, 1977.

Hoffmann, Kathryn A. "Monstrous Women, Monstrous Theorizing: Mothers, Physicians, and les Esprits Animaux." *Papers on French Seventeenth-Century Literature* 24.47 (1997): 537–53.

Homer. *Iliad*. Trans. Richmond Lattimore. Chicago: U of Chicago P, 1951.

Hooke, Robert. *Micrographia: Or Some Physiological Descriptions of Minute Bodies made by Magnifying Glasses with Observations and Inquiries thereupon*. London: Jo. Martyn and Ja. Allestry, Printers to the Royal Society, 1665.

Horkheimer, Max, and Theodor W. Adorno. *Dialectic of Enlightenment*. Trans. John Cumming. New York: Herder and Herder, 1972.

Huet, Marie-Hélène. "Monstrous Imagination: Progeny as Art in French Classicism." *Critical Inquiry* 17.4 (1991): 718–40.

Hutton, Patrick H., et al. *Historical Reflexions* [Issue on Vico] 22.3 (1996).

Ihde, Don. *Technology and the Lifeworld from Garden to Earth*. Bloomington and Indianapolis: Indiana UP, 1990.

Imperato, Francesco. *Intorno a diverse cose naturali [opera non meno curiosa, che utile, e necessaria à Professori della natural Filosofia]*. Naples: Nella Stamperia di Egidio Longo, 1628.

Impey, Oliver, and Arthur Macgregor, eds. *The Origins of Museums: The Cabinet of Curiosities in Sixteenth- and Seventeenth-Century Europe*. Oxford: Clarendon Press, 1985.

Ingegneri, Monsignor Giovanni. *Fisionomia Naturale* [Nella quale con ragioni tolte dalla Filosofia, dalla Medicina, & dalla Anatomia, si dimostra, come dalle parti del corpo humano, per la sua naturale complessione, si possa agevolmente conietturare quali siano l'inclinationi de gli huomini]. Padua/Venice: Pietro Paolo Tozzi, 1616/1615.

Jones, Richard Foster. *The Seventeenth Century*. Stanford, Calif.: Stanford UP, 1951.

Kappler, Claude. *Monstres, demons et merveilles à la fin de Moyen Age*. Paris: Payot, 1980.

Kestler, Johann Stephan. *Physiologia Kircheriana Experimentalis*. Amsterdam, 1680.

Kircher, Athanasius. *Arca Noê*. Amsterdam, 1675.

——. *Ars Magna Lucis et Umbrae*. Rome, 1646.

——. *China Illustrata*. Trans. Van Tuyl. Muskogee, Oklahoma: Indiana UP, 1987.

——. *Ioco-Seriorum Naturae et Artis sive Magiae Naturalis*. Frankfurt, 1672.

——. *Mundus subterraneus*. 2 vols. Amsterdam: Janssonio-Waesbergiana, 1678.

——. *Musurgia Universalis*. Rome, 1650 ed. Hildesheim: Georg Olms, 1970.

——. *Oedipus Aegyptiacus*. 3 vols. Rome, 1652–54.

——. *Turris Babel*. Amsterdam, 1679.

——. *The Vulcano's: or, Burning and Fire-vomiting Mountains*. London: J. Darby, 1669.

Koyré, Alexandre. *Etudes d'histoire de la pensée scientifique*. Paris: Gallimard, 1973.

Krautheimer, Richard. *The Rome of Alexander VII*. Princeton, N.J.: Princeton UP, 1985.

Kuhn, Thomas. *The Structure of Scientific Revolutions*. Ed. Otto Neurath. Chicago: U of Chicago P, 1970.

La Mettrie, Julien Offray de, ed. *L'Homme-machine*. 1774 ed. 2 vols. Hildesheim: Georg Olms Verlag, 1970.

Lascault, Gilbert. *Le Monstre dans l'art occidental: un problème esthétique*. Paris: Klincksieck, 1973.

Legati, Lorenzo. *Museo Cospiano* [Annesso a quello del famoso Ulisse Aldrovandi E donato

alla sua Patria dall'Illustrissimo Signor Ferdinando Cospi Patrizio di Bologna e Sena-
tore, ecc.]. Bologna, 1677.

Le Goff, Jacques, ed. *La Nouvelle histoire*. Retz CEPL (1978) ed. Paris: Editions Complexe,
1988.

Lenormant, François. *La Divination et la science des présages chez les Chaldéens*. Paris: Maisonneuve et Cie., 1875.

Levine, Joseph M. "Giambattista Vico and the Quarrel between the Ancients and the Moderns." *Journal of the History of Ideas* 52.1 (1991): 55–80.

Levinson, Paul. "Toy, Mirror, and Art: The Metamorphosis of Technological Culture." In
Philosophy, Technology, and Human Affairs, ed. Larry Hickman. College Station, Tex.:
IBIS Press, 1985.

Liceti, Fortunio. *De la Nature, des causes, des différences des monstres*. Trans. François Hous-
say. Paris: Editions Hippocrate, 1937.

——. *De Monstrorum caussis, natura, et differentiis*. Padua, 1634.

Lilla, Mark. *G. B. Vico: The Making of an Anti-Modern*. Cambridge, Mass.: Harvard UP,
1993.

Lloyd, G. E. R., ed. *Hippocratic Writings*. London: Penguin Books, 1983.

Lovejoy, Arthur. *The Great Chain of Being*. Cambridge, Mass.: Harvard UP, 1964.

Lovejoy, Arthur, and G. Boas. *Primitivism and Related Ideas in Antiquity*. Baltimore: Johns
Hopkins UP, 1935.

Lucarelli, Giuliano. *Gli Orti Oricellai. Epilogo della politica fiorentina del Quattrocento e inizio
del pensiero politico moderno*. Lucca: Maria Pacini Fazzi Editore, 1979.

Lucentini, Franco. "Automatopoietica." *Almanacco letterario Bompiani* (1962): 152–58.

Luck, George. *Arcana Mundi: Magic and the Occult in the Greek and Roman Worlds*. Baltimore: Johns Hopkins UP, 1985.

Lucretius. *On the Nature of the Universe*. Trans. R. E. Latham. Middlesex: Penguin Books,
1988.

Lugli, Adalgisa. *Naturalia et Mirabilia. Il collezionismo enciclopedico nelle Wunderkammern
d'Europa*. Milan: Mazzotta, 1983.

Luhmann, Niklas. *The Differentiation of Society*. New York: Columbia UP, 1982.

Lycosthenes [Wolffhart], Conradus. *Prodigiorum ac ostentorum chronicon*. Basle, 1557.

Magalotti, Lorenzo. *Saggi di naturali esperienze*. Milan: Longhanesi, 1976.

Mali, Joseph. *The Rehabilitation of Myth: Vico's New Science*. Cambridge: Cambridge UP,
1992.

Malinowski, Bronislaw. *Magic, Science, and Religion and Other Essays*. Garden City, N.Y.:
Doubleday Anchor Books, 1948.

Mandeville, Bernard. *A Treatise of the Hypochondriack and Hysterick Diseases (1730)*. 2d ed. of
1711 original ed. Delmar, N.Y.: Scholars' Facsimiles and Reprints, 1976.

Manesson-Mallet, Allain. *Description de l'univers*. Paris, 1683.

Maravall, José Antonio. *Culture of the Baroque: Analysis of a Historical Structure*. Ed. Wlad
Godzich and Jochen Schulte-Sasse. Trans. Terry Cochran. Vol. 25. Minneapolis: U of
Minnesota P, 1986.

Marchi, Armando. "Il Seicento en enfer. La narrative libertina del '600 italiana." *Letteratura
italiana* 2 (1984): 351–67.

Marcuse, Herbert. *One Dimensional Man*. Boston, Mass.: Beacon Press, 1964.

Marino, Giovan Battista. *La Galeria*. 2 vols. Ed. M. Pieri. Padua: Livana, 1979.

Marino, G. B., et al. *Epistolario*. Bari: Laterza, 1912.

Massey, Lyle. "Anamorphosis through Descartes or Perspective Gone Awry." *Renaissance
Quarterly* 50.4 (1997): 1148–89.

Martin, Dr. Ernest. *Histoire des monstres depuis l'antiquité jusqu'à nos jours.* Paris: C. Reinwald, 1880.

Mazzotta, Giuseppe F. *The New Map of the World: The Poetic Philosophy of Giambattista Vico.* Princeton, N.J.: Princeton UP, 1999.

Miner, Robert C. "Verum-factum and Practical Wisdom in the Early Writings of Giambattista Vico." *Journal of the History of Ideas* 59.1 (1998): 53–74.

Mitcham, Carl, ed. *Philosophy and Technology II.* Dordrecht: D. Reidel, 1986.

Molinari, Cesare. *Le Nozze degli dèi.* Rome: Bulzoni, 1968.

More, St. Thomas. *Utopia.* Ed. Edward Surtz. New Haven: Yale UP, 1972.

Morpurgo-Tagliabue, Guido. *Anatomia del Barocco.* Palermo: Aesthetica edizioni, 1987.

New Oxford Annotated Bible with the Apocrypha. Revised Standard Version. New York: Oxford UP, 1977.

Niccoli, Ottavia. *Profeti e popolo nell'Italia del Rinascimento.* Rome-Bari: Laterza, 1987.

——. *Prophecy and People in Renaissance Italy.* Trans. Lydia G. Cochrane. Princeton, N.J.: Princeton UP, 1990.

Nicéron, F. Jean François. *La Perspective curieuse ou magie artificièle des effets merveilleux.* Paris: Pierre Billaine, 1638.

Nicolini, Fausto. *Commento storico alla Seconda Scienza Nuova.* 2 vols. Rome: Edizioni di storia e letteratura, 1978.

——. *La giovinezza di Giambattista Vico.* Bologna: Mulino, 1992.

——. *Sulla vita civile, letteraria e religiosa napoletana alla fine del seicento.* Naples: Tipografia Sangiovanni, 1929.

Nocciera, G. *Il segno barocco.* Rome: Bulzoni, 1983.

Oldani, Louis J., and Victor R. Yanitelli. "Jesuit Theater in Italy: Its Entrances and Exit." *Italica* 76.1 (1999): 18–32.

Olmi, Giuseppe. "Science-Honour-Metaphor: Italian cabinets of the sixteenth and seventeenth centuries." *The Origins of Museums.* Ed. Oliver Impey and Arthur Macgregor. Oxford: Clarendon Press, 1985.

Ovid. *Metamorphoses.* Trans. Rolfe Humphries. Bloomington: Indiana UP, 1955.

Pagel, Walter. "The Reaction to Aristotle in Seventeenth-Century Biological Thought." In *Science, Medicine, and History: Essays on the Evolution of Scientific Thought and Medical Practice,* ed. E. Ashworth Underwood. 2 vols. London: Oxford UP, 1953.

Pagel, W. and F. N. L. Poynter. "Harvey's Doctrine in Italy . . ." *Bulletin of the History of Medicine* 34.5 (1960): 419–29.

Pallavicino, Sforza. *Opere edite ed inedite del Cardinale Sforza Pallavicino.* Rome: Salviucci, 1844.

——. *Trattato dello stile e del dialogo (1662).* Reprinted in *Trattatisti e narratori del Seicento,* ed. Ezio Raimondi. Milan/Naples: Riccardo Ricciardi Editore, 1960.

Paré, Ambroise. *Des Monstres et prodiges.* Ed. Jean Céard. Geneva: Librairie Droz, 1971.

——. *On Monsters and Marvels.* Trans. Janis L. Pallister. Chicago: U of Chicago P, 1982.

Parel, Anthony J. *The Machiavellian Cosmos.* New Haven: Yale UP, 1992.

Park, Katherine, and Lorraine Daston. "Unnatural Conceptions: The Study of Monsters in France and England." *Past & Present* 92 (1981): 20–54.

Parker, Geoffrey, and Lesley M. Smith, eds. *The General Crisis of the Seventeenth Century.* London: Routledge and Kegan Paul, 1978.

Passerano, Alberto Radicati. *Recueil de pièces curieuses sur les matières les plus intéressantes.* Rotterdam, 1736.

Pattaro, Sandra Tugnoli. *Metodo e sistema delle scienze nel pensiero di Ulisse Aldrovandi.* Bologna: Clueb, 1981.

Pennington, D. H. *Europe in the Seventeenth Century.* London: Longman, 1989.

Peregrini, Matteo. *Delle acutezze, che altrimenti spiriti, vivezze, e concetti, volgarmente si appellano, trattato.* Genoa: C. Ferroni, 1639.

Phinella, Philippi (Filippo Finella). *Naturali Phisonomia planetaria: Revolutionibus annorum: Duobus Conceptionis Figiuris; et de connexione inter eas, et Figuram Coelestem NEC NON Speculum Astronomicum.* Naples: Per Aegidium Longo, 1648.

Pisetzky, Rosita Levy. "Il gusto barocco nel costume italiano del '600." *Studi secenteschi* 2 (1961): 61–94.

Pizzamiglio, Gilberto. "Il Catalogo ritrovato. Giovanartico di Porcìa e la storiografia letteraria nel primo settecento." In *Tra storia e simbolo.* Vol. 46 of *Biblioteca di "Lettere Italiane" Studi e Testi.* Florence: Olschki, 1994.

Plato. *The Collected Dialogues of Plato.* Ed. E. Hamilton and C. Huntington. Princeton: Princeton UP, 1985.

Pliny. *Natural History Libri III–VII.* Ed. G. P. Goold. Trans. H. Rackham. Loeb Classical Library ed. Vol. 2. Cambridge, Mass.: Harvard UP, 1989.

Pomian, Krzysztof. *Collectionneurs, amateurs et curieux. Paris—Venise.* Paris: Gallimard, 1987.

Pompa, Leon. *Vico: A Study of the "New Science."* Cambridge: Cambridge UP, 1990.

Pomponazzi [1462–1525], Pietro. "Apologia (1517)/De naturalium effectum admirandorum, seu De incantationibus (1520)." In *The Renaissance Philosophy of Man,* ed. Ernst Cassirer, Paul Oskar Kristeller, and John Herman Randall. Trans. William Henry Hay II. Chicago: U of Chicago P, 1948.

Porta, Giambattista della. *De Humana Phisiognomia.* Naples, 1586.

——. *La Fisonomia dell'huomo et la celeste di Gio: Battista Dalla Porta* [Libri Sei. Tradotti di latino in Volgare, & hora in questa Nova Forma, & ultima Editione ricorreta, & postovi le Figure di Rame à proprio suoi luoghi, & cavate le vere effiggie dalle Medaglie, e Marmi, che nell'altre stampe non sono. Con la Fisonomia Naturale di Monsignor Giovanni Ingegneri di Polemone, & Adamantio]. Venice: Presso Sebastian Combi and Gio: LaNoù. Alla Minerva, 1652.

——. [J.-B. Porta]. *La Magie naturelle.* Rouen (1631) ed. Vendée: Editions de la Maisnie, 1975.

——. *Natural Magick.* Ed. Derek J. Price. London, 1658 ed. New York: Basic Books, 1957.

Porta, Gio. Battista Della, and Giovanni Ingegneri. *Della Celeste Fisonomia di Gio.* [Battista Della Porta Napoletano, Libri Sei nei quali ributtata la vanità dell'astrologia giudiciaria, si dà maniera di esattamente conoscere per via delle cause naturali tutto quello, che l'aspetto, la presenza, & le fattezze de gl'huomini possono fisicamente significare, e promettere. Opera nova, & piena di dotta curiosità]. In the same volume: *Fisionomia Naturale* di Monsignor Giovanni Ingegneri Vescovo di Capo D'Istria. Padua/Venice: Pietro Paolo Tozzi, 1616/1615.

Porta, Giovan Battista Della, et al. *La Fisonomia dell'huomo et la celeste* [Del Signor Giovan Battista Dalla Porta. Libri Sei. Tradotti di Latino in volgare, & hora in questa Settima, & ultima Impressione ricorretta, & postovi le figure à propri suoi luoghi. Con la Fisonomia Naturale di Monsignor Giovanni Ingegnieri, di Polemone, di Adamantio, & il Discorso di Livio Agrippa sopra la Natura, & Complessione, Humana, con il Trattato di Nei di Lodovico Settali Gentilhuomo Milanese. Aggiontovi di nuovo la Metoposcopia di Ciro Spontone]. Venice: Nicolò Pezzana, 1668.

Praz, Mario. *Studi sul concettismo.* Florence: Sansoni, 1946.

——. *Studies in Seventeenth-Century Imagery.* Rome: Edizioni di storia e letteratura, 1964.

Rabb, Theodore K. *The Struggle for Stability in Early Modern Europe.* Oxford: Oxford UP, 1975.

Ragionieri, Giovanna. *Il giardino storico italiano.* Siena and San Quirico d'Orcia: Olschki, 1978.

Raimondi, Ezio. *Anatomie secentesche.* Pisa: Nistri-Lischi, 1966.

——. "La nuova scienza e la 'visione degli oggetti.'" In *Rappresentazione artistica e rappresentazione scientifica nel "secolo dei lumi,"* ed. Vittore Branca. Venice: Sansoni, 1970.

——. *Letteratura barocca.* Florence: Olschki, 1961.

——, ed. *Trattatisti e narratori del Seicento.* Milan/Naples: Riccardo Ricciardi Editore, 1960.

Ratto, Franco. "Vico Revisited." *Forum Italicum* 31.1 (1997): 231–50.

Raymond, Janice G. *Women as Wombs: Reproductive Technologies and the Battle over Women's Freedom.* San Francisco: Harper, 1993.

Redondi, Pietro. *Galileo Heretic.* Trans. Raymond Rosenthal. Princeton, N.J.: Princeton UP, 1987.

Reeves, Eileen. *Painting the Heavens: Art and Science in the Age of Galileo.* Princeton, N.J.: Princeton UP, 1997.

Riccio, Agostino Del. "Del giardino di un rè." In *Il Giardino storico italiano,* ed. Heikamp Detlef. Florence: Olschki, 1981.

Ripa, Cesare. *Iconologia.* 3d. ed. Rome, 1603.

——. *Iconologia.* Piero Buscaroli ed. Milan: TEA, 1992.

Rossi, Paolo. *I filosofi e le macchine (1400–1700).* Milan: Feltrinelli, 1962.

——. "Le similitudini, le analogie, le articolazioni della natura." *Intersezioni* 4.2 (1984).

Royal Society of London. *Philosophical Transactions.* London, 1665.

Sahlins, Marshall. *Islands of History.* Chicago: U of Chicago P, 1985.

Saint-Hilaire, Isidore Geoffroy. *Traité de Tératologie* [Histoire générale et particulière des anomalies de l'organisation chez l'homme et les animaux, ouvrage comprenant des recherches sur les caractères, la classification, l'influence physiologique et pathologique, les rapports généraux, les lois et les cause, des monstruosités, des variétés et vices de conformation, ou]. Paris: J.-B. Baillière, 1832.

Scarabelli, Pietro Francesco, and Paolo Maria Terzago. *Museo, ò Galleria* [Adunata dal sapere, e dallo studio del Sig. Canonico Manfredo Settala nobile milanese descritta in Latino deal Sig. Dott. Fis. Colleg. Paolo Maria Terzago. Et poi in italiano dal Sig. Pietro Francesco Scarabelli Dottor Fisico di Voghera, & dal medemo accresciuta. Et hora ristampata con l'Aggiunta di diverse cose poste nel fine de medemi Capi dell'Opra ed.]. Tortona, 1677.

Schmidt, Charles B., ed. *The Cambridge History of Renaissance Philosophy.* Cambridge: Cambridge UP, 1988.

Schmitt, Charles B., and C. Webster. "Harvey and M. A. Severino: A Neglected Medical Relationship." *Bulletin of the History of Medicine* 14.1 (1971): 49–75.

Schott, Gaspar. *Magia Optica.* Frankfurt, 1677.

——. *Magia Universalis Naturae et Artis.* Bamberg, 1676.

——. *Mechanica hydraulico-pneumatica.* Frankfurt, 1657.

——. *Physica Curiosa, sive Mirabilia Naturae et Artis Libris XII.* Herbipoli, 1667.

——. *Technica Curiosa.* Nuremberg, 1664.

Schupbach, William. "Cabinets of Curiosities in Academic Institutions." In *The Origins of Museums,* ed. Oliver Impey and Arthur MacGregor. Oxford: Clarendon Press, 1985.

Scot, Reginald. *The Discoverie of Witchcraft.* 1930 ed. New York: Dover Publications, 1972.

Scotti, Mario, ed. *Prose scelte di Daniello Bartoli e Paolo Segneri.* Turin: U.T.E.T., 1967.

Selig, Karl Ludwig. "Gracián and Alciato's 'Emblemata.'" *Comparative Literature* 8 (1956): 1–11.

Serres, Michel. "C'était avant l'Exposition (Universelle)." In *Junggesellenmaschinen/Les machines célibataires,* ed. Jean Clair and Harald Szeeman. Venice: Alfieri, 1975.

Settala, Lodovico. *Della Ragion di Stato* [libri sette di Lodovico Settala All'Illustrissimo, & Eccellentissimo Signore Don Emanuelle de Fonseca e Zugniga, Conte di Monterrey, e di Fontes, del Consiglio di Stato di Sua Maestà Cattolica, & suo Presidente del Sopremo d'Italia]. Milan: Appresso Gio. Battista Bidelli, 1627.

Séverin, Marc Aurèle. *Corps de Médecine.* [De la médecine efficace ou la manière de guérir les plus grandes & dangereuses maladies tant du dedans que du dehors, par le fer & par le feu. Divisée en III. livres. par Marc Aurèle Séverin professeur en anatomie & chirurgie en l'Académie Royale de Naples. Et traduite nouvellement de Latin en François, avec les tables des chapitres et matières.] Ed. Théophile Bonet. Geneva: Pour Iean Anthoine Chovët, 1679.

Severini, Marcus Aurelius. *Zootomia Democritzea* [Idest, Anatome Generalis totius animantium Opificii, libris quinque distincta, quorum seriem sequens facies delineabit. Opus, quod omnes omnium bonarum artium Studiosos, nedum Professores anatomicos decet. Marci Aurelii Severini, Thurii Tharsiensis, Philosphi Medici, Primarii Regio in Auditorio Neapolitano Anatomes & Chirurgiae Professoris. Venite & vidcte opera Domini, quae posuit (ex animantibus) prodigia super terram, Psalm. 46.v.9]. Noribergae: Literis Endterianis, 1645.

Shumaker, Wayne. *Natural Magic and Modern Science.* Vol. 63. Binghamton, N.Y.: Medieval and Renaissance Texts and Studies, 1989.

Singer, Charles, et al., eds. *A History of Technology.* Oxford: Clarendon Press, 1954–84.

Skorupsky, John. *Symbol and Theory: A Philosophical Study of Theories of Religion in Social Anthropology.* Cambridge: Cambridge UP, 1976.

Slaughter, M. M. *Universal Languages and Scientific Taxonomy in the Seventeenth Century.* Cambridge: Cambridge UP, 1982.

Spence, Jonathan D. *The Memory Palace of Matteo Ricci.* New York: Viking Penguin, 1984.

Staum, Martin. "Physiognomy and Phrenology at the Paris Athenée." *Journal of the History of Ideas* 56.3 (1995): 443–62.

Stelluti, Francesco. *Della Fisonomia di tutto il corpo humano del S. Gio.Battista Porta* [Acc. Linceo Libri Quattro Ne'quali si tratta di quanto intorno a questa materia n'hanno i Greci, Latini, e gli Arabi scritto. Hora brevemente in tavole sinottiche ridotta et ordinata. Da Francesco Stelluti. Acc. Linceo da Fabriano all Eminentissimo et Reverendissimo Sig. Cardinale Frac. Barberino]. Rome: Vitale Mascardi, 1637.

Stone, Harold Samuel. *Vico's Cultural History: The Production and Transmission of Ideas in Naples, 1685–1750.* New York: E. J. Brill, 1997.

Tagliacozzo, Giorgio. *The Arbor Scientiae Reconceived and the History of Vico's Resurrection.* Atlantic Highlands, N.J.: Humanities Press, 1993.

——. "Unity of Knowledge: From Speculation to Science." *New Vico Studies* 14 (1996): 139.

Tambiah, Stanley Jeyaraja. *Magic, Science, Religion, and the Scope of Rationality.* Cambridge: Cambridge UP, 1990.

Taruffi, Cesare. *Storia della Teratologia.* Bologna: Regia Tipografia, 1881.

Tesauro, Emanuele. *Il Cannocchiale Aristotelico.* Zavatta, 1670 ed. Berlin: Verlag Gehlen, 1968.

Tesauro, Lodovico. *Esamina del co. A. dell'Arca intorno alle Ragioni del co. L. Tesauro in difesa di un sonetto del Cav. Marino.* Bologna, 1614.

Thomas, Keith. *Religion and the Decline of Magic.* New York: Scribner's, [1971].

Thorndike, Lynn. *A History of Magic and Experimental Science.* New York: Columbia UP, 1951.

Topsell, Edward. *The History of Four-Footed Beasts and Serpents and Insects.* Facsimile, 1658 ed. New York: Da Capo, 1967.

Torrini, Maurizio. *Tommaso Cornelio e la ricostruzione della scienza.* Studi Vichiani. Naples: Guida, 1977.

Traupman, John C. *The New College Latin and English Dictionary.* New York: Bantam Books, 1966.

Trevor-Roper, H. R. *The European Witch-Craze of the Sixteenth and Seventeenth Centuries and Other Essays.* New York: Harper and Row, 1968.

Tribby, Jay. "Body/Building: Living the Museum Life in Early Modern Europe." *Rhetorica* 10.2 (1992): 139–64.

———. "Cooking Clio and Cleo: Eloquence and Experiment." *Journal of History of Ideas* 52.3 (1991): 417–40.

Troncarelli, Fabio. "La Paura dell'Idra. Kircher e la peste di Roma." In *Enciclopedismo in Roma barocca,* ed. Maristella Casciato et al. Venice: Marsilio Editori, 1986.

Turner, Victor W. "Divination as a Phase in a Social Process." In *Reader in Comparative Religion,* ed. W. A. Lessa and Evon Z. Vogt. New York: Harper and Row, 1968.

Valton, Edmond. *Les Monstres dans l'art.* Paris, 1905.

Varchi, Benedetto. *Lezzioni di M. Benedetto Varchi.* Florence: Filippo Giunti, 1590.

Veeser, Aram, ed. *The New Historicism.* New York: Routledge, 1989.

Verene, Donald Phillip. *The New Art of Autobiography: An Essay on the Life of Giambattista Vico, Written by Himself.* Oxford: Clarendon Press; New York: Oxford UP, 1991.

Verene, Molly Black. *Vico: A Bibliography of Works in English from 1884 to 1994.* Bowling Green, Ohio: Philosophy Documentation, 1994.

Vico, Giambattista. *L'Autobiografia.* Ed. F. Nicolini. Bologna: Mulino, 1992.

———. *L'Autobiografia, il carteggio e le poesie varie.* Ed. B. Croce and F. Nicolini. Vol. 5. Bari: Laterza, 1929.

———. *Autobiografia. Seguita da una scelta di lettere, orazioni e rime.* Ed. Mario Fubini. Turin: Einaudi, 1977.

———. *The Autobiography of Giambattista Vico.* Trans. M. H. Fisch and T. G. Bergin. Ithaca: Cornell UP, 1983.

———. *De antiquissima italorum sapientia: indici e ristampa anastatica.* Ed. Giovanni Adamo. Florence: Olschki, 1998.

———. *The New Science.* Trans. T. G. Bergin and M. H. Fisch. Ithaca: Cornell UP, 1986.

———. *On Humanistic Education (Six Inaugural Orations, 1699–1707).* Trans. Giorgio A. Pinton and Arthur W. Shippee. Ithaca: Cornell UP, 1993.

———. *On the Most Ancient Wisdom of the Italians.* Trans. L. M. Palmer. Ithaca: Cornell UP, 1988.

———. *Opere.* Ed. Fausto Nicolini. In *La letteratura italiana: storia e testi.* Vol. 43. Milan: Ricciardi, 1953.

———. *Opere giuridiche. Il diritto universale.* Ed. Paolo Cristofolini. Florence: Sansoni, 1974.

———. *Opere. La Scienza Nuova prima.* Ed. Fausto Nicolini. Anastatic copy of 1931 ed. Vol. 3. Bari: Laterza, 1968.

———. *Opere. Scritti vari e pagine sparse.* Ed. Fausto Nicolini. Vol. 7. Bari: Laterza, 1940.

———. *Principi di una scienza nuova.* Ed. Marco Veneziani. Florence: Olschki, 1997.

———. *Scritti storici.* Ed. Fausto Nicolini. Vol. 6. Bari: Laterza, 1939.

Vigliani, Luigi. "Emanuele Tesauro e la sua opera storiografica." *Fonti e studi di storia fossanese* 14 (1936): 207–77.

Vovelle, Michel. *Histoires figurales.* Florence: Usher, 1989.

Walker, D. P. *Spiritual and Demonic Magic from Ficino to Campanella.* Ed. G. Bing. Vol. 22. Nendeln/Liechtenstein: Klaus Reprint, 1976.

Webster, Charles, ed. *The Intellectual Revolution of the Seventeenth Century.* London: Routledge and Kegan Paul, 1974.

Wellek, René. "The Concept of Baroque in Literary Scholarship," *Journal of Aesthetics and Art Criticism* 5.2 (1946): 77–109.

White, Hayden. *Tropics of Discourse: Essays in Cultural Criticism.* Baltimore: Johns Hopkins UP, 1978.

White, Lynn. "Cultural Climates and Technological Advance in the Middle Ages." *Viator* 2 (1971): 171–201.

Williams, David. *Deformed Discourse: The Function of the Monster in Medieval Thought and Literature.* Montreal: McGill-Queen's UP, 1996.

Willis, Thomas. *Two Discourses concerning The Soul of Brutes.* Trans. S. Pordage. Facsimile reproduction of 1683 ed. Gainesville, Fl.: Scholars' Facsimiles and Reprints, 1971.

Winner, Langdon. *Autonomous Technology: Technics-out-of-Control as a Theme in Political Thought.* Cambridge, Mass.: MIT P, 1977.

Wittkower, Rudolf. "Marvels of the East: A Study in the History of Monsters." *Journal of the Warburg and Courtauld Institutes* 5 (1942): 159–97.

Wolff, Etienne. *La Science des monstres.* Paris: Gallimard, 1948.

Wölfflin, Heinrich. *Principles of Art History.* Trans. M. D. Hottinger. G. Bell and Sons (1932) ed. New York: Dover, 1950.

Yates, Frances A. *The Art of Memory.* Chicago: U of Chicago P, 1966.

——. *Giordano Bruno and the Hermetic Tradition.* Chicago: U of Chicago P, 1964.

Zambelli, Paola. *Le problème de la magie naturelle à la Renaissance.* Warsaw, Poland: Zaklad Narodowy Imienia Ossolinskich Wydawnictwo Polskiej Akademii Nauk, 1972.

Zonca. *Novo teatro di machine.* Padua, 1607.

INDEX

Winner, Langdon, 92, 94–95

Wit and witticisms. *See* Rhetorical figures

Wittgenstein, Ludwig: use of a cognitive schema, xi, 219n. 2

Women as monsters, 8, 104, 105, 107, 109 (fig. 14), 113–17, 217. *See also* Sirens

Wonder (admiration, *meraviglia*), 5, 11, 13, 17, 82, 84, 156, 181, 187–217, 247n. 7. *See also* Astonishment; Baroque aesthetics; Curiosity; Stupor

Wonders of nature. *See* Nature: wonders of

Zakiya Hanafi is an independent scholar
(formerly Assistant Professor of French and Italian
at the University of Washington at Seattle).

Library of Congress Cataloging-in-Publication Data
Hanafi, Zakiya
The monster in the machine : magic, medicine, and the
marvelous in the time of the scientific revolution / Zakiya Hanafi.
Includes bibliographical references and index/
ISBN 0-8223-2536-5 (alk. paper)—
ISBN 0-8223-2568-3 (pbk. : alk. paper)
1. Monsters—Italy—History—17 century. 2. Body
Human—Italy—Philosophy—History—17th century. I. Title.
QM691 .H257 2001 306.4'5—dc21 00-035458